PLANT GENOMES: METHODS FOR GENETIC AND PHYSICAL MAPPING

Plant Genomes:
Methods for Genetic
and Physical Mapping

Edited by

J.S. BECKMANN

*Department of Plant Genetics and Breeding, Agricultural Research Organization,
The Volcani Center, Bet-Dagan, Israel*

and

T.C. OSBORN

Department of Agronomy, University of Wisconsin, Madison, WI, USA

Chapters indicated with an asterisk in the table of contents were first published
in the Plant Molecular Biology Manual, as Supplement 5,
ISBN 90-247-3633-1.

SPRINGER–SCIENCE+BUSINESS MEDIA, B.V.

Library of Congress Cataloging-in-Publication Data

Plant genomes : methods for genetic and physical mapping / edited by
  Jacques S. Beckmann and Thomas C. Osborn.
     p.   cm.
  Includes bibliographical references and index.
  ISBN 978-94-010-5077-7   ISBN 978-94-011-2442-3 (eBook)
  DOI 10.1007/978-94-011-2442-3
  1. Plant genome mapping.   I. Beckmann, Jacques S.   II. Osborn,
Thomas C.
  QK981.45.P53   1992
  581.5'87322--dc20   92-2530

ISBN 978-94-010-5077-7

Table of contents

Preface

JACQUES S. BECKMANN & THOMAS C. OSBORN

Extraordinary progress has been made in the analyses of the genetic structures of higher eukaryotic genomes. Only ten years elapsed between the initial proposals to use molecular DNA markers for the generation of a complete linkage map of the human genome [5, 17] and the first description of a 10 centimorgan map of one of its chromosomes [22], soon to be followed by others. The availability of molecular DNA markers, henceforth called genomic markers [for a review of their properties see 1, 2, 20], represents a milestone in genetics by providing the capacity for complete genetic coverage of all genomes.

It is important to remember that the nature of the DNA polymorphism or of the specific method used to uncover it can be quite different for different marker loci. The genetic variation detected can be a result of a simple point mutation, a DNA insertion/deletion event, or a change in repeat copy number at some hypervariable DNA [11] or microsatellite [21] motif. Currently, the methods of detection can involve use of restriction endonucleases, nucleic acid hybridization, or DNA sequence amplification. Each of these sources of variation and methods of detection can have utility for different applications. Furthermore, new approaches for the detection of DNA polymorphism are constantly emerging. The primary concern here is that the monitored polymorphism defines a genetic marker 'useful' for the desired application.

Parallel to this enhanced genetic capability, the physical mapping and sequencing of entire genomes is, as a result of recent technological advances, slowly and gradually also turning into reality. The comparison of the genetic maps with the physical maps will be necessary to elucidate the physical/functional relationships of the different genomic parts. In other words, the analyses of different variant types and of the resulting phenotypic nuances will lead to a clearer understanding of what gene does what, where, when, why and how. Indeed, the basic rationale for mapping a genome is the understanding of the simple as well as intricate biological processes, the identification of the chromosomal coordinates and molecular cloning of the genetic loci involved in these processes. Genetic mapping is often a first, and obligatory, step toward such ends (a procedure frequently referred to as reverse genetics), while physical mapping can be an essential adjunct, providing access to defined genomic regions.

In a nutshell, the main application of genomic markers is the mapping, diagnosis, understanding, manipulation and eventual molecular cloning of loci of medical, biologic or economic interest. These achievements are likely to have

J.S. Beckmann and T.C. Osborn (eds) Plant Genomes: Methods for Genetic and Physical Mapping, vii–xi.

a practical impact not only in basic science, e.g., in the elucidation of the basic processes involved in a particular pathway or in retracing the history of plant speciation, but also in breeding practice (see, for instance, the discussions by Weller and by Knapp et al. in this book).

In agriculture this translates into the exploitation of this knowledge for further genetic improvement. It should be remembered that plant breeders are principally concerned with phenotypic traits germane to the genetic improvement of a crop species or variety. Moreover, traits of agronomic interest are often polygenic, i.e., phenotypic expression of the trait is determined by variation at a number of chromosomal loci (the polygenes or quantitative trait loci, QTL) [8]. The elucidation of the genetic basis of polygenic inheritance and the subsequent efficient utilization of this knowledge in breeding schemes require the availability of an adequate network of genetic markers spanning the entire genome.

The interest in polygenic or multifactorial inheritance is not limited to agricultural species. But in this context, animal and plant studies offer a significant advantage over human genetics, insofar that populations, matings and, in some instances, the phasing of markers can be controlled. Thus, the capacity to handle large pedigrees, more or less designed at will, should enable (i) the examination of the inheritance of both the simple as well as complex polygenic traits, (ii) the identification of the individual Mendelian entities involved in these traits by association with closely linked genetic markers, and (iii) the use of these markers as tags in breeding programs or as reference points for the eventual cloning of these loci. This, in conjunction with transgenic methodologies, is bound to disclose an enormous amount of information on the factors underlying quantitative inheritance.

Hence, the growing interest in mapping plant genomes is not surprising. Detailed genetic maps already exist for a number of plants (for example, *Arabidopsis* [6], *Brassica* [16, 18], maize [see 10], potato [4, 7] and tomato [19]). This list is bound to grow as more and more laboratories join in this endeavor. The purpose of the present manual is to provide new investigators in this area with an introduction and a basic practical state-of-the-art description of how to proceed. These are covered in a series of chapters by Kidwell and Osborn, Bernatzky and Schilling, Palmer, and Shattuck-Eidens *et al.*

This manual is different from many others, in that it does not attempt to give a comprehensive coverage of all the techniques involved. The reader is advised, instead, to refer to the classical manuals, such as Sambrook *et al.* [15] or Gelvin and Schilperoort [9], for basic procedures. Moreover, the general area covered in this book's chapters is, *per se*, a new and rapidly evolving one. Consequently, several older as well as more recent techniques are not, or only marginally, discussed here. Among these one can cite the development of RAPD markers [23], the use of the polymerase chain reaction (PCR) to target onto sequence tagged microsatellite sites [2, 13, 21], the development of automated assays for genotyping, the utilization of non-radioactive probes, *in situ* hybridization, use of particular genetic stocks (addition lines, aneuploid lines, translocations, etc), and many other items.

Also, it should be clear to all that although much of the discussions in this manual focus on RFLPs, this does not imply that RFLPs are either the sole or most suitable type of markers. Each type of marker has its virtues and drawbacks. As an illustration, RAPD offer a very elegant, fast and relatively inexpensive method to generate rapidly (as the name suggests) a large number of genetic markers, which can have very powerful applications, such as the study of near-isogenic material [12, 14], or of specific mutant stocks. This method has, however, important limitations, insofar as the markers uncovered are often of a dominant type and can thus be specific to the cross examined. In other words, the genetic study of another cross within the same species (e.g. of two other unrelated cultivars) might require the generation *ab initio* of a novel map.

One of the well known attractions of molecular genetics is that the same tools can be applied to many species. Hence, it is not surprising that some chapters deal with applications outside of plants or methods that are just beginning to be applied to plants. For example, methods for constructing YAC libraries and of physical mapping are covered by Albertsen *et al.*, Hauge and Goodman, and van Daelen and Zabel, whereas the lessons that can be learned from genetic linkage analyses in humans, including the convenience of use of LOD scores and the use of thresholds, are covered in the chapter by Lalouel.

Yet, in spite of this 'democratization' of genomic sciences, some important differences need to be highlighted too. In human genetics, significant LOD score thresholds were chosen as to minimize the risk of a type II error. A breeder's demarche can be very different, as he or she may not worry about selecting too much, i.e. of selecting chromosomal segments erroneously recognized as having particular favorable QTLs. In contrast, he or she might care a lot about losing segments which do carry such true loci. Hence, it might be valuable to reevaluate the optimal threshold definitions for breeding applications.

Although none of the papers in this book specifically refer to it, there is one additional crucial point worth emphasizing. Linkage analyses are particularly sensitive to errors. As there are numerous steps involved in this type of analyses, there are numerous opportunities for errors to inadvertently occur (such as the mislabelling or mixups of plants or tubes, the inappropriate interpretation of incomplete restriction digests, sample contaminations, or the erroneous interpretation or entry of genotypes in the databases). The net result of most of these errors is that an incorrect genotype will be attributed to a defined phenotype or individual. In other words, this will lead, for instance, to the creation of a recombinant where there is none. It is easy to visualize that this source of errors could have dramatic effects in a fine mapping exercise: it would tend to place a gene in an interval in which it is not. And for those interested in walking toward this gene, this could send one onto a wrong path. However, even in more coarse mapping, such errors might take their toll and may obscure linkage. This issue of errors should not be taken lightly, as by the nature of this kind of work, the data can often only be interpreted long after the plants have been sampled. It is thus important that particular attention be paid, right from the onset of the work, to follow extremely rigorous protocols, and to use automation whenever possible. Better start right now, rather than realize too late one's errors.

To sum up, the main motivation for collecting these chapters together stems from our conviction that the availability and power of these new tools, and of rapidly emerging technologies and instrumentation, will lead to a dramatic increase in the knowledge of plant genomes. Many labs are likely to gradually venture in this area and contribute to these developments. Our aim was to facilitate or encourage this transition by enabling some of today's principal actors in this field to describe in detail many of the steps involved. Hopefully, this series of chapters, however partial and incomplete, will encite others to join into this new and exciting research area of gene mapping. The task is enormous, but the rewards far outweigh it.

And if we have been successful in this endeavor, it is foremost thanks to all contributors who accepted to lay down here the basic procedures or processes involved in one or another step of the analyses. We would also like to extend our deep appreciation to the publisher, for his patience and encouragements, and to all colleagues who accepted to read and anonymously review these manuscripts, and last but not least to our families for their unselfish help and comprehension.

References

1. Beckmann JS, Soller M (1986) Restriction fragment length polymorphisms in plant genetic improvement. In: Mifflin BJ (ed) Oxford Surveys of Plant Molecular and Cell Biology, Vol. 3: 196–250. Oxford: Oxford University Press.
2. Beckmann JS, Soller M (1990) Toward a unified approach to genetic mapping of eukaryotes based on sequence tagged microsatellite sites. Bio/technology 8: 930–932.
3. Beckmann JS, Soller M (1989) Genomic genetics in plant breeding. Vortr. Pflanzenzüchtg. 16: 91–106.
4. Bonierbale MW, Plaisted PL, Tanksley SD (1988) RFLP maps based on a common set of clones reveal modes of chromosomal evolution in potato and tomato. Genetics 16: 91–106.
5. Botstein D, White RL, Skolnick M, Davis RW (1980) Construction of a genetic linkage map in man using restriction fragment length polymorphism. Am J Hum Genet 32: 314–331.
6. Chang C, Bowman JL, DeJohn AW, Lander ES, Meyerowitz EM (1988) Restriction fragment length polymorphims map for *Arabidopsis thaliana*. Proc Natl Acad Sci USA 85: 6856–6860.
7. Gebhardt C, Ritter E, Debener T, Schachtsnabel U, Walkemeier B, Uhrig H, Salamini F (1989) RFLP analysis and linkage mapping in *Solanum tuberosum*. Theor Appl Genet 78: 65–75.
8. Geldermann H (1975) Investigation on inheritance of quantitative characters in animals by gene markers. I. Methods. Theor Appl Genet 46: 300–319.
9. Gelvin S, Schilperoort R (1988) Plant Molecular Biology Manual. Dordrecht: Kluwer Academic Publishers.
10. Helentjaris T, Burr B (1989) Development and Application of Molecular Markers to Problems in Plant Genetics. Cold Spring Harbor, NY: Cold Spring Harbor Laboratory Press, 165 pp.
11. Jeffreys AJ (1987) Highly variable minisatellites and DNA fingerprints. Biochem Soc Trans 15: 309–317.

12. Klein-Lankhorst RM, Vermunt A, Zabel P (1991) Isolation of molecular markers for tomato (*L. esculentum*) using random amplified polymorphic DNA (RAPD). Theor Appl Genet 83: 108–114.
13. Litt M, Luty JA (1989) A hypervariable microsatellite revealed by *in vitro* amplification of a dinucleotide repeat within the cardiac muscle actin gene. Am J Hum Genet 44: 397–401.
14. Martin GB, Williams JGK, Tanksley SD (1991) Rapid identification of markers linked to a *Pseudomonas* resistance gene in tomato by using random primers and near-isogenic lines. Proc Natl Acad Sci USA 88: 2336–2340.
15. Sambrook J, Fritsch EF, Maniatis T (1989) Molecular Cloning: A Laboratory Manual, 2nd edition. Cold Spring Harbor, NY: Cold Spring Harbor Laboratory Press.
16. Slocum MK, Figdore SS, Kennard WC, Suzuki JY, Osborn TC (1990) Linkage arrangement of restriction fragment length polymorphism loci in *Brassica oleracea*. Theor Appl Genet 80: 57–64.
17. Solomon E, Bodmer W (1979). Evolution of sickle variant gene. Lancet i: 923.
18. Song KM, Suzuki JY, Slocum MK, Williams PH, Osborn TC (1991) A linkage map of *Brassica rapa* (syn. *campestris*) based on restriction fragment length polymorphism loci. Theor Appl Genet (in press).
19. Tanksley SD, Miller J, Paterson A, Bernatzky R (1988) Molecular mapping of plant chromosomes. In: Gustafson JF, Appels R (eds) Chromosome Structure and Function, pp. 157–173. New York: Plenum.
20. Tanksley SD, Young ND, Paterson AH, Bonierbale MW (1989) RFLP mapping in plant breeding: new tools for an old science. Bio/technology 7: 257–266.
21. Weber JL, May PE (1989) Abundant class of human DNA polymorphisms which can be typed using the polymerase chain reaction. Am J Hum Genet 44: 388–396.
22. White RL *et al.* (1990) The CEPH consortium primary linkage map of human chromosome 10. Genomics 6: 393–412.
23. Williams JGK, Kubelik AR, Livak KJ, Rafalski JA, Tingey SV (1990) DNA polymorphisms amplified by arbitrary primers are useful as genetic markers. Nucleic Acids Res 18: 6531–6535.

1. Simple plant DNA isolation procedures

KIMBERLEE K. KIDWELL & THOMAS C. OSBORN
Department of Agronomy, University of Wisconsin, Madison, WI 53706, USA

Introduction

A key step in the analysis of plant DNA restriction fragments is the purification of sufficient quantities of good-quality DNA from plant tissues. High DNA yields are often needed for studies that require analysis of several Southern blots, and the purified DNA must be free of contaminants that interfere with restriction endonuclease digestion or electrophoretic separation of DNA fragments. Contaminants, such as polysaccharides, can cause shifts in mobility during electrophoresis resulting in misinterpretation of fragment differences among genotypes. This may not be a problem when analyzing a population that is segregating for a few expected restriction fragments. However, it can be a serious problem when analyzing many unique genotypes for fragment length polymorphisms.

For large genetic studies involving analysis of many samples, DNA isolation procedures must be inexpensive and rapid. Many such procedures have been described, and several of these use an extraction buffer containing cetyltrimethylammonium bromide (CTAB) [1–4, 6]. The use of CTAB in plant DNA purification was originally described for a procedure that involved precipitating DNA by lowering the concentration of NaCl in the presence of CTAB [2]. This precipitation method presumably removes many polysaccharides which are still soluble at the lower NaCl concentration. This original procedure, which included a cesium chloride ultracentrifugation step, was shortened to a more rapid, inexpensive protocol that can be used for processing many samples simultaneously [3]. Another rapid procedure using this same CTAB extraction buffer was described by Saghai-Maroof *et al.* [4]; however, in this method DNA was precipitated only once by addition of isopropanol. This latter procedure is very simple and it has been used successfully with different plant species by many researchers.

In this chapter, we describe a simple DNA isolation procedure and modifications of this procedure involving precipitation of DNA based on lowering the NaCl concentration in the presence of CTAB. We compare purification methods for DNA yield and describe the effects of these methods on electrophoretic mobility of DNA restriction fragments.

1

J.S. Beckmann and T.C. Osborn (eds) Plant Genomes: Methods for Genetic and Physical Mapping, 1–13.

Procedures

Basic procedure

For this procedure, DNA is isolated from lyophilized tissue suspended in a buffer containing CTAB. After one chloroform extraction, the DNA is precipitated with either isopropanol or ethanol. This procedure is based on methods reported originally by Murray and Thompson [2] and then modified by Saghai-Maroof *et al.* [4].

Steps in the procedure
1. Collect several grams of young leaf tissue from healthy plants, lyophilize the samples and store in a desiccator at $-20\,^\circ$C.
2. At room temperature, grind 250–300 mg of lyophilized tissue into a fine powder and transfer the tissue into a labeled 15 or 50 ml polypropylene tube.
3. Add 5–10 ml DNA extraction buffer (approximately 1 ml per 30–50 mg tissue; the optimal ratio of tissue may vary with different plant species). Gently and thoroughly suspend the tissue in the buffer using a slow, rocking motion.
4. Incubate for 60 min at 55–60 °C with occasional mixing.
5. Add an equal volume of chloroform/isoamyl alcohol (24 : 1) and gently and thoroughly mix. Centrifuge at 1000–5000 g, 20 °C for 30–50 min.
6. Using a large-bore pipette, transfer the aqueous (upper) phase to a labeled 50 ml tube and add 2.5 volumes EtOH ($-20\,^\circ$C) or 0.6–1 volume of isopropanol ($-20\,^\circ$C). Mix gently until DNA precipitates.
7. Depending on the condition of the precipitated DNA, select the appropriate method for rinsing, drying, and redissolving the sample.

A. If long strands of DNA that wind together are visible:
1. Immediately hook out the DNA using a bent and sealed pasteur pipette (DNA will adhere to the glass).
2. Rinse the DNA in a labeled 50 ml tube containing 10–20 ml of a 75% EtOH, 10 mM ammonium acetate solution. Incubate for 20 min at room temperature with occasional swirling. Repeat two or three times.
3. Place the DNA on the side of a sterile labeled 1.5 ml microfuge tube and allow it to air-dry (10–20 min). Completely remove excess alcohol from the bottom of the tube using an aspirator. DNA also can be air-dried directly on the glass hook.
4. Dissolve the DNA in 200–800 μl sterile TE buffer (or just enough to allow it to go into solution). Centrifuge at 13 000 $\times$ μg in a microfuge for 10 min to pellet undissolved particles. Measure DNA concentration and adjust to 0.5–0.7 μg/μl.

B. If the DNA cannot be hooked:
1. Centrifuge the sample at 1000 $\times$ μg for 5 min. Gently decant the supernatant, invert the tube, and allow the pellet to air-dry for 5 min.
2. Resuspend the pellet in a small volume (1–5 ml) of 0.5 M NaCl and reprecipitate with alcohol as described in step 6. Hook, rinse, dry, and redissolve as described in steps 7.A.1–7.A.4.

3. If the DNA still cannot be hooked:
   a) Centrifuge at 1000 × μg for 5 min. Without disrupting the pellet, gently decant the supernatant or aspirate it using a sterile pasteur pipette.
   b) Rinse pellet in 10–20 ml of a 75% EtOH, 10 mM ammonium acetate solution for 10 min at room temperature with occassional swirling. Repeat two to three times or until the pellet is white. Overnight incubation may be required to remove contaminants from pellets.
   c) Centrifuge at 1000 × μg for 5 min. Decant or aspirate the supernatant and thoroughly dry the pellet. The pellet can be air-dried in the inverted tube for 30–60 min or tubes can be placed in a lyophilizer for 5–10 min.
   d) Dissolve the DNA in an appropriate volume of TE (typically 200–400 μl) and transfer to a labeled 1.5 ml microfuge tube. Centrifuge at 13 000 × μg for 10 min and measure DNA concentration.

Comments

In order to obtain high yields of intact DNA, it is important to use the youngest leaves of healthy plants (see Comparisons of procedures below). Leaves near the apical or axillary meristems are the best source of tissue. The leaf tissue can be collected into Whirlpak bags (VWR) which should be stored on ice during harvest. Fresh tissue can be lyophilized immediately or frozen at $-80\,°C$ and lyophilized later. After freeze-drying, the tissue may be stored desiccated at $-20\,°C$ for several months without loss of DNA yield or quality. For long-term storage, lyophilized tissue seems to be better preserved at $-80\,°C$.

The leaf tissue must be thoroughly ground to obtain maximum yields. Best results are obtained from samples ground immediately prior to DNA extraction. For moderate-size samples (200–300 mg), a mortar and pestle is very effective for grinding tissue. High DNA yields can also be obtained from samples ground by adding glass beads to tubes containing dried tissue and shaking for several minutes using a modified paint shaker [5]. This is especially convenient when working with large numbers of samples. When tissue supply is limited, small-size samples (less than 100 mg) can be ground directly with minimal tissue loss in a 15 ml conical centrifuge tube using a glass rod that has been shaped to precisely fit the tube bottom. Mechanical grinders are convenient for larger samples (greater than 300 mg) where maximum DNA recovery is not essential. Thorough suspension of tissue in the extraction buffer is also important for maximum yields. To aid in suspension, a spatula may be used to gently break apart clumps of moistened tissue. The solution should be mixed gently (avoid excessive shaking which could shear DNA) until the tissue is homogeneously suspended in solution.

Chloroform is used to extract soluble proteins from the sample. After this step, the sample needs to be centrifuged long enough to separate the phases and to pellet the insoluble debris between the phases. Care should be taken not to disturb this pad of debris when removing the aqueous phase.

Either ethanol or isopropanol can be used to precipitate DNA; however, isopropanol may be preferred due to the lower volume required for DNA precipitation. At low concentrations, DNA may not precipitate in sufficient quantities to allow hooking. If long strands of hookable DNA are not visible after the first alcohol precipitation, the DNA usually can be hooked after precipitating a second time from a smaller volume. This second precipitation also helps to remove contaminants from the samples.

Glass wool can also be used to precipitate and collect DNAs that are dissolved at low concentrations. Disperse a pinch of unsiliconized glass wool in a labeled 50 ml tube containing 15 ml of isopropanol (room temperature). Transfer the aqueous phase described in step 6 of the basic procedure into this tube and mix well by inverting the capped tube several times then incubate at room temperature for several minutes. The precipitated DNA strands should become visible in the glass wool. Decant the solution while holding the glass wool inside the tube with forceps. Wash the DNA/glass wool by mixing with 10 ml of 75% ethanol, 10 mM ammonium acetate as described in the basic procedure. Following the final wash, decant the solution then use forceps to squeeze as much liquid as possible from the DNA/glass wool. Adhere DNA/glass wool to the side of the tube and invert the tube for several minutes to drain off the excess alcohol. Dissolve the DNA by incubating the DNA/glass wool in 5 ml of sterile distilled water at 60 °C for 1 h or leaving it at room temperature for 3 h to overnight. After the DNA has dissolved, squeeze the excess liquid from the glass wool then discard it. Add 0.6 ml of 5 M NaCl to the DNA solution and precipitate with 2.5 vol. of ethanol (-20 °C) or 0.6–1 vol. of isopropanol (-20 °C). Rinse, dry, and redissolve the DNA as described in step 7 of the basic procedure.

Some precipitated DNA samples may take a long time to dissolve. Heating to 55–60 °C for 5 min followed by gentle shaking or leaving out overnight at room temperature may help dissolve the sample. DNA dissolved in TE buffer can be stored at 4 °C for at least several months.

In order to avoid sample mix-ups, care must be taken to ensure that all plants, bags, and tubes are labeled with the appropriate plant identification code. To avoid loss of identification by smearing ink with chloroform or alcohol, both the cap and side of each tube should be labeled using a permanent ink marker. Tubes used for alcohol precipitation can be easily washed and reused; however, we discard tubes used for the initial tissue suspension and choloform extraction because they are difficult to clean and they are weakened by the chloroform.

Modifications of the basic procedure

DNA isolated from young, healthy tissue using the basic procedure will usually be suitable for restriction enzyme digestion and Southern blot analysis. However, in cases where tissue of less than optimal condition is available, further steps may be required in order to obtain high-quality DNA. Modifications of the basic procedure involving a CTAB precipitation method [2, 3], can help to eliminate contaminants from DNA samples that interfere with electrophoretic mobility. Although the CTAB procedure seems to result in better quality DNA, lower yields are recovered using this method when compared to the basic procedure (see Comparisons of procedures for discussion). The same basic procedure described above is followed for steps 1 through 5 of the modified method, then the DNA is precipitated as described below.

CTAB precipitation
6a. Remove the aqueous phase using a large-bore pipette and place in a labeled 50 ml tube. Add a 1/10 volume of 10% CTAB solution and mix gently.
7a. Add an equal volume of chloroform/isoamyl alcohol (24 : 1) and mix gently and thoroughly. Centrifuge at $1000-5000 \times \mu g$, 20 °C for 20 min.
8a. Transfer the aqueous phase to a clean, labeled 50 ml polypropylene centrifuge tube and add an equal volume of 1% CTAB solution. Mix gently and thoroughly. Incubate at room temperature for 30 min (CTAB will precipitate at low temperatures).
9a. Typically, DNA precipitated by this method is not hookable, so centrifuge and then rinse, dry, and resuspend the pellet as described in step 7.B.3a–7.B.3d of the basic procedure.

Solutions

- DNA extraction buffer
  - 0.7 M NaCl
  - 50 mM Tris-HCl pH 8.0
  - 10 mM EDTA pH 8.0
  - 1% CTAB
  - 0.1% β-mercaptoethanol
- TE buffer
  - 10 mM Tris-HCl pH 7.5
  - 1 mM EDTA pH 8.0
- 10% CTAB
  - 10% CTAB
  - 0.7 M NaCl
- 1% CTAB
  - 1% CTAB
  - 50 mM Tris-HCl pH 8.0
  - 10 mM EDTA pH 8.0

Comparisons of procedures

The yield and quality of DNA may differ for the different procedures described in this chapter. The maturity of the tissue used for DNA extraction also may affect the yield and quality of DNA obtained from these procedures. We compared DNAs isolated from different leaf tissue samples of *Medicago sativa* and *Brassica* species using the basic and modified DNA isolation procedure described above.

Effect of tissue condition and maturity on DNA yield

Leaves of three maturity levels (determined by leaf size) were collected from greenhouse-grown plants of the following sources: *M. sativa* (2 ×), *B. rapa* (syn. *campestris*), healthy plants of *B. oleracea*, and *B. oleracea* grown under stressed conditions (densely planted and water-stressed). Equal weights of lyophilized tissue from each source were ground in a mortar and pestle to a fine powder which was then divided into equal parts. DNA was isolated from one sample of each source using the basic procedure with one isopropanol precipitation while the other sample was processed using the modified CTAB precipitation procedure. DNA concentrations were determined by fluorometry in order to compare DNA yields from the two procedures.

   In most cases, the highest DNA yields for both procedures were obtained from the samples of young leaves while the lowest yields were obtained from the more mature leaves (data not shown). The effect of maturity level on DNA yield also differed depending on the species from which the DNA was isolated. As leaf maturity increased, DNA yields from the *Brassica* species decreased more dramatically than did yields from *M. sativa*. In comparing DNA yields from healthy and stressed leaf tissue of *B. oleracea*, much lower DNA yields were obtained from stressed tissue at all levels of tissue maturity for both precipitation methods.

DNA yields using the two precipitation methods

Leaves from the same range in maturity were collected from three clonally propagated individuals of *B. oleracea* cv. Badger Inbred 16. Six hundred milligrams of lyophilized tissue from each plant were ground to a fine powder followed by DNA extraction as described in steps 1 through 5 of the basic procedure. Following step 5, the supernatant was divided into two equivalent fractions. DNA was recovered from one fraction using the basic procedure (steps 6–7.A.4) and from the other fraction by using the CTAB precipitation method (steps 6a–9a). Fluorometry was used to determine DNA concentrations of each sample. The DNA yields (μg/200 mg lyophilized tissue $\pm$ standard deviation) were substantially higher for the basic procedure (51.8 $\pm$ 4.6) than for the procedure including the CTAB precipitation (33.5 $\pm$ 1.6).

Quality of DNAs isolated using the different precipitation methods

The relative qualities of DNAs isolated from young leaf tissue using the basic procedure and the modified procedure including CTAB precipitation were determined by comparing electrophoretic separations of DNAs digested with a restriction enzyme. Three quantities of DNA (2, 5 and 10 μg) isolated from young, healthy leaf tissue of *M. sativa*, *B. oleracea* and *B. rapa* were digested with *Hin*d III using 3 units of enzyme per μg of DNA in a 30 μl reaction. Reactions were incubated at 37 °C for 5 h followed by separation of fragments in a 0.8% agarose gel using 40 mM Tris-acetate, 1 mM EDTA as the electrophoresis buffer.

DNAs isolated from all species tested using both methods were fully digested by *Hin*d III at all DNA concentrations (results shown for *M. sativa* in Fig. 1A). The ethidium bromide staining intensity increased proportionally with increased DNA concentration for both isolation procedures (Fig. 1A, lanes 1, 2 and 3 and lanes 4, 5 and 6) and the gel separation of restriction fragments from isopropanol and CTAB precipitation appeared identical.

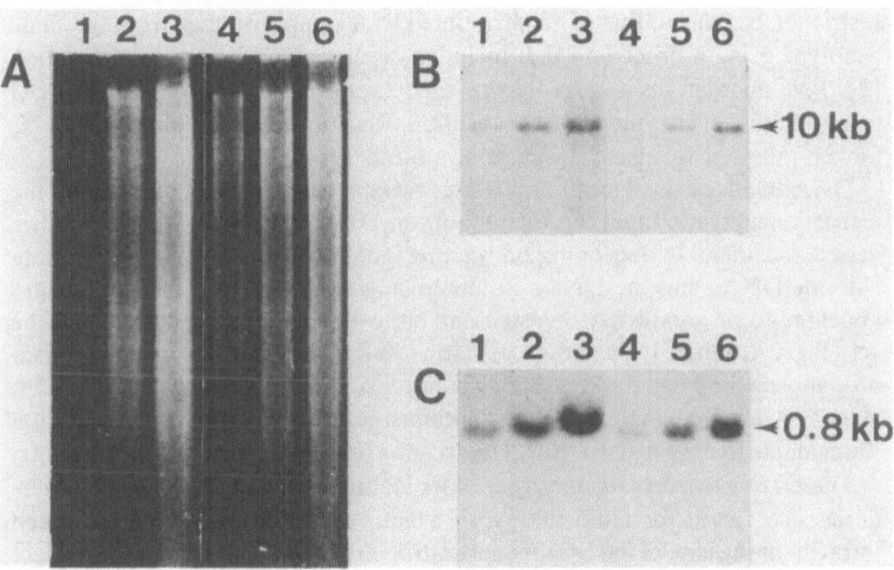

Fig. 1. Comparison of DNA quality in samples isolated using the basic procedure and the CTAB modified procedure (see Procedures section). A. Photograph of ethidium bromide-stained agarose gel containing various quantities of *Hin*d III-digested 2 × *Medicago sativa* DNA isolated from young leaves. Lanes 1 through 3 contain 2, 5, and 10 μg, respectively, of DNA isolated using the basic procedure. Lanes 4 through 6 contain 2, 5, and 10 μg of DNA isolated using the CTAB modified procedure. B and C. Autoradiograph of a Southern blot of the gel in A hybridized to a ^{32}P-labeled low-copy nuclear DNA sequence from *M. sativa*. Homologous fragments of 10 kb (B) and 0.8 kb (C) are shown.

In order to verify the uniform mobility of identical restriction fragments in DNA samples obtained using the isopropanol or the CTAB procedures, the gel in Fig. 1A was Southern blotted and hybridized to a low-copy, anonymous genomic DNA sequence. In the resulting autoradiograph, a 10 kb (Fig. 1B) and a 0.8 kb (Fig. 1C) band are visible in each lane. The hybridization signal intensity increased with increasing DNA concentration for both isolation procedures. The 10 kb band had a uniform shape and mobility at all concentrations for both DNA preparations (Fig. 1B). Although the 0.8 kb band (Fig. 1C) had a more U-shaped appearance at the higher DNA concentrations for the samples prepared using the basic procedure, no obvious shifts in fragment mobility were observed.

The qualities of DNAs isolated from less than optimal tissue were compared for the two isolation procedures using DNAs from three individuals of *B. oleracea* cv. Badger Inbred 16. The tissue used for DNA isolation included young and old leaves collected from plants that had been maintained in the greenhouse for several months. DNAs were extracted from equivalent amounts of lyophilized tissue from three plants using the two precipitation methods as described previously (see *DNA yields using the two precipitation methods*). The relative quality of the DNAs isolated using the two procedures was determined by comparing the electrophoretic separation of these DNAs digested with a restriction enzyme to that of an ultra-pure DNA sample isolated from the same genotype using a procedure that included cesium chloride ultracentrifugation [2]. Five micrograms of DNA from each sample were digested with *Eco*RI using 15 units of enzyme per 20 μl reaction. Reactions were incubated at 37 °C for 5 h followed by electrophoresis in a 0.8% agarose gel.

DNAs isolated using both procedures were at least partially digested by the restriction enzyme; however, the amount of DNA that entered the gel matrix varied considerably depending on the precipitation method used. The cesium chloride DNAs and the DNAs isolated using the modified CTAB procedure appeared to be completely digested and almost all of the samples entered the gel (Fig. 2A, lanes 1, 5, and 6–9). However, DNAs isolated using the isopropanol method had darkly staining bands visible in the sample wells (Fig. 2A, lanes 2–4, arrow) which may have contained undigested DNAs or DNAs that were unable to migrate into the gel matrix due to the presence of contaminants.

The DNA samples isolated from older leaf tissue using the different procedures also varied for uniformity with which restriction fragments separated throughout the lane of the agarose gel matrix. Differences in lane uniformity are probably due to differences in the purity of isolated DNA samples. Restriction fragments of the highest-quality DNAs, i.e., samples isolated using cesium chloride ultracentrifugation, showed very uniform separation in the gel lanes (Fig. 2A, lanes 1, 5, and 9). Nonuniform fragment separation was evident in DNAs isolated using the basic isopropanol method (Fig. 2A, lanes 2–4). However, samples isolated using the modified CTAB procedure separated uniformly (Fig. 2A, lanes 6–8) and were nearly indistinguishable from the samples isolated using cesium chloride ultracentrifugation (Fig. 2A, lanes 1, 5,

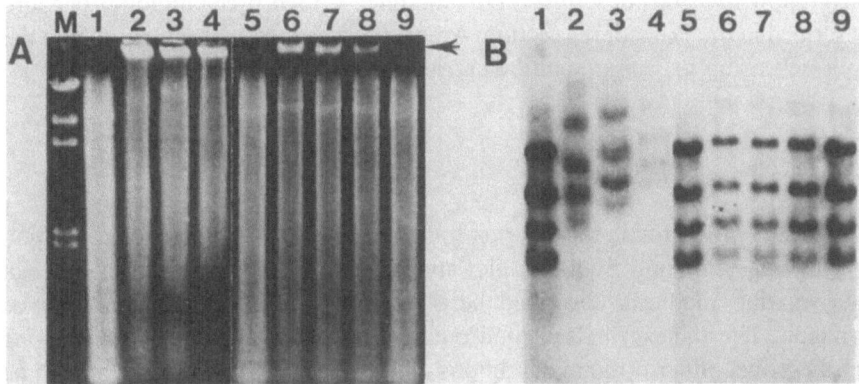

Fig. 2. Comparison of DNA quality in samples isolated from older leaf tissue using the basic isopropanol and the CTAB precipitation procedures. A. Photograph of ethidium bromide-stained agarose gel containing 5 μg of *Eco*RI-digested DNAs isolated from individuals of *Brassica oleracea* cv. Badger Inbred 16 using the two isolation methods. Lanes 2 through 4 contain samples isolated using the basic procedure while lanes 6 through 8 contain DNAs isolated using the modified CTAB procedure. Cesium chloride-purified DNA from this same genotype is shown in lanes 1, 5, and 9. Lane M contains *Hin*d III-digested lambda DNA. B. Autoradiograph of a Southern blot of gel in A hybridized to a ^{32}P-labeled *B. napus* cruciferin probe, pC1.

and 9). Thus, precipitation of DNA by lowering the NaCl concentration in the presence of CTAB probably eliminates some contaminants in older leaf tissue samples that co-precipitate with DNA when alcohol is used.

In order to determine the effect of differences in lane uniformity on mobility of identical restriction fragments in the DNAs isolated using the different procedures, the gel in Fig. 2A was Southern blotted and hybridized to pC1, a cDNA clone of the cruciferin seed protein gene from *B. napus* [7]. In the resulting autoradiograph, a number of bands are visible in each lane (Fig. 2B). Since all DNA samples were isolated from plants with identical genotypes, each lane should show the same banding pattern with restriction fragments in identical positions. However, there are obvious differences between lanes in the position of restriction fragments that are related to the appearance of the samples in the agarose gel. The position of DNA fragments from the basic isopropanol samples (Fig. 2B, lanes 2–4) are shifted upward and highly variable when compared to the cesium chloride-purified DNA (Fig. 2B, lanes 1, 5, and 9) and the modified CTAB-precipitated samples (Fig. 2B, lanes 6–8). DNA fragments from the cesium chloride-purified sample and all CTAB-precipitated samples had nearly identical mobilities suggesting that the modified CTAB procedure effectively removes contaminants from DNA samples isolated from older leaf tissues.

DNA samples which are prepared using the basic isopropanol precipitation method and then found to have shifted mobility after electrophoresis can be cleaned up by using several steps of the modified CTAB method. Begin the procedure by adding an equal volume of 2 × extraction buffer to the sample then follow the steps described in the modified CTAB precipitation procedure described in the Procedures section. Adjust the volumes accordingly so that the entire clean-up procedure can be carried out in 1.5 ml microfuge tubes.

Conclusions

The results presented in this chapter indicate that DNA suitable for restriction enzyme digestion and Southern blot analysis can be recovered using either of the isolation methods described here when tissue of optimal condition is available. The highest yields of good-quality DNAs are obtained by using young leaves grown under optimal conditions, and therefore, extreme care should be taken to grow healthy plants and to harvest the youngest leaf tissue. When using optimum leaf tissue, the basic isopropanol procedure gives good results and is the preferable method because of its simplicity and high DNA yields. We have used this method successfully to isolate DNA from *Brassica*, *Lycopersicon*, *Medicago*, and *Phaseolus* species. When starting with older or stressed leaf tissue, however, fragment mobility shifts may be observed in samples isolated using the basic procedure. Samples isolated using the modified CTAB precipitation procedure do not exhibit fragment mobility shifts for any of the tissues we have tested, and thus, if tissue of less than optimal quality must be used for DNA isolation, it may be beneficial to include the CTAB precipitation steps.

Acknowledgements

The authors thank Keming Song and Brian Diers for technical assistance in testing the protocols and Martha Crouch for providing the pC1 probe. This work was supported by a Pioneer Graduate Fellowship (to K.K.), the USDA/CRGO (grant 89–37140–4636) and the College of Agricultural and Life Sciences, University of Wisconsin-Madison.

13

References

1. Keim P, Olson TC, Shoemaker RC (1988) A rapid protocol for isolating soybean DNA. Soybean Genet Newsl 15: 150–152.
2. Murray MG, Thompson WF (1980) Rapid isolation of high molecular weight plant DNA. Nucleic Acids Res 8: 4321–4325.
3. Rogers SO, Bendich AJ (1985) Extraction of DNA from milligram amounts of fresh, herbarium and mummified plant tissues. Plant Mol Biol 5: 69–76.
4. Saghai-Maroof MA, Soliman KA, Jorgensen, RA, Allard, RW (1984) Ribosomal DNA spacer-length polymorphisms in barley: Mendelian inheritance, chromosomal location, and population dynamics. Proc Natl Acad Sci USA 81: 8014–8018.
5. Tai TH, Tanksley SD (1990) A rapid and inexpensive method for isolation of total DNA from dehydrated plant tissue. Plant Mol Biol Rep 8: 297–303.
6. Webb DM, Knapp SJ (1990) DNA extraction from a previously recalcitrant plant genus. Plant Mol Biol Rep 8: 180–185.
7. Simon AE, Tenbarge KM, Scofield SR, Finkelstein RR, Crouch ML (1985) Nucleotide sequence of a cDNA clone of *Brassica napus* 12S storage protein shows homology with legumin from *Pisum sativum*. Plant Mol Biol 5: 191–201.

2. Methods of Southern blotting and hybridization

ROBERT BERNATZKY & ANGELA SCHILLING
Department of Plant and Soil Sciences, University of Massachusetts, Amherst, MA 01003, USA

Introduction

At present, the technique that remains central to RFLP analysis is Southern blotting and hybridization [15]. Briefly, the procedure involves the enzymatic cleavage of DNA with restriction endonucleases, the separation of the resultant fragments by electrophoresis through an agarose gel and the transfer of the fragments from the gel to a membrane that binds nucleic acids. The fragments are transferred in a single-stranded state by treating the gel with alkali which disrupts the hydrogen bonding between the native double-stranded DNA. The membrane-bound DNA is then hybridized in solution to specific radiolabelled DNA sequences (probes) that anneal to complementary sequences on the membrane. Through autoradiography, one or a few fragments can be visualized separately from the very large number of fragments that make up a complex genome.

Since the introduction of the technique by Southern, procedural improvements have been developed. The basic process of DNA digestion and electrophoresis remains the same. Major advances have been made in the efficient transfer of fragments from the gel and in the membranes used to fix the DNA. There have also been, and continue to be, improvements made in the preparation of probes and hybridization of the probes to the membranes.

The large size of some restriction fragments after endonuclease digestion makes transfer of these fragments out of the gel by capillary action difficult to accomplish. Treatment of the gel with acid results in partial depurination of the DNA and subsequent cleavage at the depurination sites when the gel is treated in alkali [17]. These smaller fragments then move more easily out of the gel during the blotting process.

Nitrocellulose membranes have been widely used for Southern blotting. However, the brittle nature of nitrocellulose makes multiple reuse of the membranes difficult. The development of nylon-based membranes significantly improved both the blotting and hybridization process [12]. These membranes are more durable than nitrocellulose. In addition, they have a higher binding capacity for nucleic acids, particularly smaller fragments, and bind both double- and single-stranded DNA (nitrocellulose binds single-stranded DNA only). There is also the advantage that the DNA can be transferred from the gels under low-ionic-strength, alkaline conditions. Blotting in alkali eliminates the need for a neutralization step and ensures that the DNA is transferred in a

15

J.S. Beckmann and T.C. Osborn (eds) Plant Genomes: Methods for Genetic and Physical Mapping, 15–33.

single-stranded state. Alkali also promotes the covalent fixation of the DNA to the nylon membranes. This results in a stronger retention of the bound DNA and allows the filters to be re-used many times [12].

The hybridization process has been advanced by the addition of dextran sulfate to the hybridization solution. It has been shown that anionic dextran polymers accelerate the reassociation of DNA in solution [18] as well as in the two-phase system of Southern hybridization [17]. The increase in rate of reassociation has been attributed to the concentrating effect of the dextran sulfate, that is, exclusion of the DNA from the volume occupied by the polymer [12, 13].

Probe preparation has been improved by the technique of random primer synthesis [4]. The labelling method of nick translation [13], although suitable for many applications, has disadvantages. The reaction temperature must be maintained at 15 °C and the nicking of the template DNA with DNase I must be controlled. Radiolabelling DNA by random primers can achieve specific activities of 10^9 dpm/μg template DNA in as little as one hour, can be done at room temperature, and can allow for extended reaction time without sacrificing probe quality.

The primary goal of the Southern blotting and hybridization procedures is to produce clean, resolvable restriction fragment patterns that are reproducible under specific conditions. If these procedures are applied to a long-term mapping project then an important concern is how many times the membranes can be used. Most of the work that goes into a mapping project involves DNA isolation, digestion, gel electrophoresis and Southern blotting. The more times that a given membrane can be used to map new loci, the less time and expense is needed for additional DNA, restriction enzymes, agarose and blotting membrane. There is a range of qualities of membranes manufactured by different companies, and each type needs to be tested under individual laboratory conditions for their ability to bind and maintain DNA through multiple usage.

We have tested two hybridization conditions and two stripping procedures (i.e. removal of bound probe for reuse of membrane with a different probe) for their effects on the longevity of nylon membrane Southern blots. *Brassica rapa* 'Candle' DNA (1–2 μg) was digested with *Pst* I and subjected to electrophoresis and blotting as described later under Procedures. Eight replicate membranes were divided into two groups and each group was either hybridized at 68 °C in a buffer containing 5 × SSC, 50 mM sodium phosphate pH 7.2, 5 × Denhardt's solution, 2.5 mM EDTA, 0.6% SDS, 0.1 mg/ml denatured salmon sperm DNA and 5% dextran sulfate or hybridized at 42 °C in a buffer composed of 5 × SSC, 50 mM sodium phosphate pH 7.0, 10 × Denhardt's solution, 1.0% SDS, 2 mg/ml salmon sperm DNA, and 50% formamide. With regard to membrane longevity, we believed that the most significant variables between these hybridization protocols would be temperature, and the presence of formamide. The probes employed were derived from random, low-copy, *Pst* I genomic fragments of 'Candle' cloned into pUC 18. Hybridization occurred for 16 h. Membrane washing procedures (described later) were the same for both groups.

After exposing the membranes to film and evaluating relative signal strength, each group of four membranes was divided again and the probe was stripped off using two procedures, either an alkali treatment or a high temperature/low salt buffer. The alkali procedure employed 0.1 M NaOH at 42 °C for 10 min followed by a 15 min wash in 0.1 M Tris pH 7.5, 0.1× SSC, and 0.1% SDS and then a 15 min wash in 0.1× SSC and 0.1% SDS at room temperature. The high temperature/low salt treatment consisted of two 10 min washes of 0.1× SSC and 0.1% SDS at 80 °C.

Each membrane went through three cycles of hybridization (using different probes) and stripping. This entire experiment was then repeated using another set of eight filters. Two findings were clear: 1) the hybridization at 68 °C without formamide was a superior system, producing signals that were five to ten times stronger than those obtained from hybridizing at 42 °C in the presence of formamide and 2) the two different probe stripping procedures affected the longevity of the membranes similarly, that is, membranes that were stripped differently gave similar signals in subsequent hybridizations. The stronger hybridization signals obtained at 68 °C could be due to a combination of the increased rate of hybridization at the higher temperature and the presence of dextran sulfate. The effects of dextran sulfate in the 42 °C formamide buffer was not tested.

To test whether the different hybridization conditions had any effect on the longevity of the membranes, sets of membranes that had been hybridized three times under the two conditions were combined and hybridized under the 68 °C procedure. In this case, all filters gave equivalent signals, suggesting that the hybridization conditions did not affect the life of the membranes.

We conclude that the longevity of the Southern blots was not affected by the hybridization or probe stripping procedures. Our present opinion is that the source and type of membrane may be the most significant factor conditioning the ability to reuse membranes. However, all of these results must be considered in light of the specific mechanical procedures used (i.e. length of hybridization time, hybridization containers, washing procedures, etc.) and should not be considered universal. Since each step in these procedures, from DNA isolation to membrane stripping, affects subsequent procedures, a variety of membranes should be tested under individual laboratory procedures in order to optimize the use of the membranes.

Procedures

The following protocols are samples of the many versions of these basic procedures. These protocols are presently used in our lab. However, we expect changes as new techniques are developed and incorporated into the Southern blotting and hybridization process. Many of the protocols are modifications of those described in Bernatzky and Tanksley [3] and Bernatzky [2]. Additional useful references are Hames and Higgins [6] and Sambrook *et al*. [14].

Restriction enzyme digestion of DNA

The choice of which restriction enzymes to use depends primarily on cost and the ability of the enzymes to detect polymorphisms. Enzymes that have 6 bp recognition sites are most commonly used since they generate restriction fragments within a size distribution that is easily manipulated by standard Southern analysis (i.e. roughly 300 bp to 20 kb). We routinely use *Dra* I, *Eco* RI, *Eco* RV, *Hin*d III, *Pst* I, and *Xba* I. The amount of enzyme used should be in the range of three to ten units per µg of DNA and the reaction should proceed for a few hours to overnight. The unit definition is the amount of enzyme needed to digest 1 µg of standard, purified DNA (usually lambda phage) in one hour. Excess enzyme is used for plant DNA to ensure complete digestion. The minimum number of units required per µg of plant DNA needs to be empirically determined. The specificity of cleavage is affected by glycerol concentrations greater than 5% for many enzymes. Since enzymes are stored in 50% glycerol, the reaction volume limits the maximum amount of units that can be used per digestion.

The majority of enzymes have optimum reaction temperatures of 37 °C, although some enzymes that are isolated from thermophilic organisms will have higher temperature optimums (e.g. 60 °C). Reaction conditions and other specifics are usually supplied by the manufacturer. Enzymes are stored in a −20 °C frost free freezer and should always be kept on ice during use.

The two principal variables in restriction enzyme digestion of DNA is the concentration of the DNA sample to be digested and the maximum sample volume that can be loaded on a gel. The amount of DNA to be digested should reflect the size of the genome under investigation. Large genomes, such as those found in the grasses, may require 10 µg or more per gel lane in order to obtain a significant autoradiographic signal in a reasonable period of time. Smaller genomes require less DNA per lane to obtain equivalent signals. We routinely use 1−2 µg of *Brassica rapa* DNA, which has a relatively small genome size of 0.6 pg per haploid genome [19].

Most gels accommodate between 15 and 50 µl sample sizes. If the DNA is too dilute to digest the proper amount of DNA in the maximum volume allowed by the gel, then the sample needs to be concentrated (i.e. by ethanol precipitation) either before or after digestion. Volumes that are smaller than the maximum allowed by the gel can be accommodated by topping off the sample after placing it in the well. Buffers for digestion are usually supplied by the manufacturer as a 10× stock or can be made according to specifications supplied with the enzyme. The following describes digestion of 2 μg of DNA for a final gel loading volume of 35 μl.

Steps in the procedure
1. Place 2 μg of DNA in a microcentrifuge tube in combination with water to a final volume of 26 μl.
2. Add 3 μl of 10× stock of appropriate buffer and mix well (but do not vortex in order to avoid shearing the DNA).
3. Remove enzyme from freezer and place on ice. Transfer 1 μl of enzyme (10 U/μl) to DNA solution and mix very well. Incomplete mixing will lead to partial digestion.
4. Place the sample at the appropriate incubation temperature for 3 h to overnight.
5. Stop the reaction with 5 μl of gel loading buffer and mix well.

   Digested DNA samples can be run immediately or stored for a short period in a refrigerator or for a longer period in a freezer.

Gel electrophoresis

DNA fragments are most commonly separated by electrophoresis through an agarose gel matrix in the presence of a buffer. Due to the common negatively charged phosphorous backbone, all fragments have the same charge-to-mass ratio. Therefore separation of fragments is based strictly on size. The distance that the DNA must run through the gel depends on the degree of separation required so that a polymorphism can be visualized. Agarose concentrations are usually in the range of 0.7–1.2%. The separation of large fragments is accomplished in the lower percent range, although gels become difficult to handle as the percentage of agarose is decreased. The rate at which DNA travels through the gel is roughly proportional to the voltage applied and inversely proportional to the ionic strength of the buffer. At a given voltage, decreasing the buffer strength by one half will cause the DNA to carry more current and move through the gel at twice the speed. Running the gels too quickly through low ionic strength buffers can sometimes lead to distortion and excessive heating. Low ionic strength buffers also become more quickly depleted and the gel becomes distorted due to a rise in pH at the anode. The protocol described below is designed for a 20 × 20 × 0.75 cm gel to be run long (16 h, overnight) in a relatively high ionic strength buffer. Other schedules can be accommodated by manipulations of buffer strength and voltage.

A lane containing molecular weight size standards is useful for determining DNA fragment lengths as well as orienting the blot. A common marker is lambda phage DNA that has been digested with *Hin*d III. Since the marker lane will be probed along with single- or low-copy sequences, it is easiest to use a small amount of marker (10–20 ng) in order to avoid excessive signal from the marker lane.

Steps in the procedure
1. Place 2.7 g of agarose into a flask that contains 300 ml of neutral electro-phoresis buffer (NEB) for a 0.9% gel. The agarose can be melted either on a hot plate with stirring or in a microwave. The flask should be covered to prevent excessive evaporation and the need to adjust volume. The agarose solution should boil for at least 30 s to ensure that all particles are dissolved.
2. The agarose should be cooled to 65 °C before pouring in order to avoid deformation of the plastic gel casting tray. This can be accomplished in an ice bath with stirring.
3. Many gel casting trays are open-ended and the ends need to be taped securely to prevent leaks. Pour the agarose solution into the tray and slowly place the well-forming 'comb' into it. Avoid getting agarose on the sides of the teeth of the comb above the surface of the gel, as this will lead to streaking of the DNA sample.
4. Allow the gel to harden for at least 30 min. Carefully remove the comb and the tape.
5. Place the gel in its casting tray into an electrophoresis unit. Pour enough NEB into the unit so that the buffer just comes to the surface of the gel but does not cover it.

22

6. Load the digested DNA samples into the wells using a microliter pipettor. Once they are all loaded, top off the wells with NEB so that each well is full. Run the samples into the gel at approximately 45 V, 75 mA for about 20 min or until all of the bromophenol blue tracking dye has left the well and entered the gel.
7. Use a small amount of buffer to wet the surface of the gel and place an acetate plastic sheet on the surface to prevent the gel from drying out during the run. Avoid trapping bubbles in the wells.
8. Run the gel for about 16 h or until the dye front has reached 12 cm.
9. Always wear gloves when handling ethidium bromide. Stain the gel in a tray of distilled water that contains approximately 100 μg of ethidium bromide (one drop of a 10 mg/ml stock solution per liter of water). Stain the gel for 15 min, rinse the gel in distilled water and destain the gel for 15 min in water. A thin sheet of plexiglass under the gel makes the gel easier to handle and transport.
10. Wearing UV-blocking glasses or face shield, observe the gel under short-wave UV light to determine if the DNA has digested and the gel has run properly (depending on the enzyme, this usually means a somewhat uniform distribution of the DNA from the wells through the tracking dye). Highly repeated sequences (generally nuclear or chloroplast DNA) may be visible as discrete bands. Photograph the gel if desired.

Southern blotting

A critical step in the Southern blotting procedure is the treatment of the gel in dilute HCl. Insufficient treatment will result in poor transfer of large DNA fragments and overtreatment will produce very small fragments that will not bind well to the membrane or will not hybridize well to probes.

The simplest DNA transfer method relies only on the liquid in the gel to mobilize the DNA and can be referred to as 'dry blotting'. Alternatively, an apparatus can be assembled that will allow a continual flow of buffer through the gel to ensure that all possible DNA has been transferred out of the gel. In either case, staining the gel after transfer will determine if the transfer is complete. We usually use the dry blotting method and rarely have trouble with incomplete transfer. The transfer is usually complete in 4 h, although it may be more convenient to transfer overnight. The dimensions of the blot depend on how far the gel was run and the upper limit depends on the size of the container used for hybridization. In the method to be presented the container is 14.5 × 20 cm. The following protocol describes an alkaline transfer process.

Steps in the procedure

1. Trim the gel to 14.5 cm from the wells.
2. Cut a piece of blotting membrane to the dimensions of the gel (handle the membrane with gloves). Write in one corner of the membrane with pencil in order to orient and identify the blot.
3. Cut 3 sheets of 3MM paper to the dimensions of the gel.
4. Soak the gel in a tray with 1 liter 0.25 M HCl for 10 min with gentle shaking. The tracking dye should turn yellow at this time.
5. Pour off the acid, rinse the gel with water and soak the gel in 0.4 M NaOH for 25 min.
6. While the gel is being treated in the acid and the base prepare the blotting set up as follows.

Wet the membrane in 0.4 M NaOH by carefully laying it on the surface of the solution so that it wets evenly, and then submerge it. Soak 2 of the 3MM sheets in the base as well.

Dry blot. Prepare a stack of single-fold paper towels about 4 cm high. After treating the gel, place it on a piece of plastic wrap or acetate sheet on a smooth surface. Briefly blot the surface of the gel with the third dry sheet of 3MM paper to remove excess buffer. Carefully lay down the blotting membrane on the gel without trapping any bubbles between the blot and the gel. Similarly lay down the 2 wet sheets of 3MM paper. Place the stack of paper towels on top of the 3MM papers making sure that there is no space between them. Place a light weight such as a thin plexiglass sheet on top and blot for 4–6 h or overnight.

Flow-through blot. Set up a sheet of glass or plastic so that it is above a tray of 0.4 M NaOH. Place 2 sheets of 3MM paper on the glass so that paper hangs down

2 opposite sides and touches the base solution. Wet the paper thoroughly with the base and remove any bubbles that may be trapped between the paper and the glass by rolling over it with a glass pipette. After treating the gel, place it on the blotting apparatus. Frame the gel with 3 cm wide strips of parafilm. Trim a stack of paper towels to the dimensions of the gel. Briefly blot the surface of the gel with the third dry sheet of 3MM paper to remove excess buffer. Carefully lay down the blotting membrane on the gel without trapping any bubbles between the blot and the gel. Similarly lay down the 2 wet sheets of 3MM paper. Place the stack of paper towels on top of the 3MM papers so that the towels cover the gel exactly and will not droop over the sides when wet. This, along with the parafilm frame, will prevent the buffer from being drawn around the gel rather than through it. Place a light weight on top, such as a thin plexiglass sheet, and blot overnight. The wet paper towels can be replaced with dry ones in order to extend the blotting time.

7. Wash the membranes after blotting in a solution of $2 \times$ SSC, 0.1% SDS at room temperature, followed by a second wash in $0.2 \times$ SSC, 0.1% SDS at room temperature. Air-dry the membranes on 3MM paper and store under low humidity at $4\,^{\circ}$C.

Probe preparation

The following is a protocol for labelling DNA probes by random priming [4]. Probe preparation by nick-translation can be found in Rigby *et al.* [13] or Sambrook *et al.* [14].

We have used commercially available kits for random primer labelling as well as a kit developed in our lab according to a modified method of Sambrook *et al.* [14]. A comparison of components revealed a much larger amount of random hexamer primers in the commercial kit. We have looked at the time course of labelling of the two kits using a number of different template DNAs and have concluded that the higher concentration of primers leads to a much more rapid rate of incorporation. In fact, maximum specific activity was achieved with the higher level of primers at the first time point of 30 min and did not change appreciably after 2 h. However, the reaction with the lower level of primers produced the same specific activity after 1 to 2 h. We also examined the products of the types of reactions on alkaline agarose gels [14] and found that the reaction with the higher level of primers produced relatively shorter labelled fragments, presumably because the higher concentration of primers leads to an increased number of primed sites. Using the two types of probes in replicate Southern hybridizations did not give appreciably different qualities of hybridization, although this was not exhaustively tested.

We prefer to label insert DNA rather than whole plasmid to reduce the amount of non-hybridizing radiolabelled DNA and to avoid any potential binding of the plasmid to non-target sequences on the blot. Inserts can be prepared according to methods outlined in Sambrook *et al.* [14] or Tautz and Renz [16].

The labelling reaction and subsequent Southern hybridization involves the use of ^{32}P and all proper precautions, such as wearing gloves and protective clothing and shielding behind plexiglass, should be taken.

Steps in the procedure
1. Combine 25–50 ng template DNA with 60 ng of random hexamer primers in a total volume of 15 μl.
2. Heat the mixture at 100 °C for 5 min either in a heat block or a boiling water bath. Cool the sample on ice.
3. Add the following:
   - 5 μl 10× labelling buffer
   - 6 μl of a solution of dATP, dGTP, and dTTP each at 5 mM
   - 20 μl H$_2$O
   - 1 μl (3 U) Klenow enzyme
   - 5 μl (50 μCi) [α-^{32}P]dCTP

   Mix the components well and incubate at room temperature for 2 or more hours.
4. Stop the the reaction with 5 μl of column-dye buffer.
5. Unincorporated dNTPs can be removed by passing the sample over a small column of Sephadex G-50 (see below). The fraction that contains the blue dextran dye (the first dye to come off the column) contains the labelled DNA.
6. Mix the sample well. Place 1% or less of the sample into a microcentrifuge tube

and place in a scintillation vial. Count the sample in a liquid scintillation counter to monitor the success of the reaction. The cpm can be converted to dpm based on the cpm of known amounts of ^{32}P. The specific activity should be between 10^8 and 10^9 dpm/μg of template DNA.

7. The labelled DNA can be used immediately or stored at $-20\,^\circ$C.

Column preparation

1. Plug a 1 ml syringe with a small amount of cotton or silanized glass wool.
2. Fill the column with a slurry of Sephadex G-50 in column buffer.

The columns can be used in two ways:

a) Spin the column at low speed for 30 s in a microcentrifuge tube in a microcentri-fuge. This will remove most of the column buffer. After applying the labelling reaction mix to the top of the column, add an additional 50 μl of column buffer to the column and spin again at low speed for 30 s. The labelled DNA along with the blue dextran dye will be in the microcentrifuge tube. The unincorporated dXTPs and the bromophenol blue dye will still be in the column.

b) Apply the reaction mix to the column. Carefully add column buffer and allow the sample to drip through. Collect the fraction that contains the blue dextran dye.

Southern hybridization

Anderson and Young [1] provide a lengthy discussion of membrane hybridization and I will mention only a few of the facts here. Filter hybridization depends on two processes: diffusion of the probe to the membrane and hybridization of the probe to the membrane-bound sequences. When the concentration of target sequence on the membrane is very low, as in Southern analysis of low-copy nuclear sequences, the hybridization reaction, and not diffusion, is the rate-limiting step. As stated previously, dextran sulfate will increase the rate of hybridization, although it is expensive and can sometimes lead to high backgrounds.

The melting temperature (T_m) is defined as that temperature at which DNA strands are half denatured or dissociated. T_m is dependent on base composition, ionic strength, and the presence of denaturing agents such as formamide. Hybridizations are usually carried out at 20–25 °C below T_m. For well-matched hybrids, the incubation temperature is 68 °C for aqueous buffers, and 42 °C for buffers that contain 50% formamide. For every 1% of mismatch, T_m is reduced by roughly 1 °C, depending on %GC and distribution of the mismatch. Mismatched hybrids can either be stabilized or dissociated through manipulations of salt and temperature. As the ionic strength of the hybridization or wash buffer is reduced, the stringency of hybrid association is increased. Washing membranes in low-ionic-strength buffer can be used to distinguish between members of multigene families that reside at different loci, making otherwise complex patterns locus-specific [11].

The hybridization protocol is designed for hard plastic hybridization boxes (15 × 22 cm Rubbermaid drawer organizers) that incubate in a hot air incubator at 68 °C. The box hybridization method is cleaner and more convenient than the standard heat-sealed plastic hybridization bags. Six to eight membranes can be hybridized in a single box. The boxes can be stacked to accommodate more than one hybridization experiment. It is often convenient to pre-hybridize the filters while the probe is being prepared.

Shaking the boxes during prehybridization and hybridization is not required but is important when washing the excess probe from the membranes. Efficient removal of probe during the wash is accomplished by placing pieces of nylon window screen between the membranes in the wash buffer.

Incubators that are specifically designed for Southern hybridization have been developed and are being tested.

Steps in the procedure
1. Remove hybridization buffer from the refrigerator and warm it in order to get the SDS back in solution. Prepare 50 ml of buffer if one membrane is to be hybridized and 25 ml or less for each additional membrane.
2. Melt frozen, denatured salmon sperm DNA and add it to the buffer for a final concentration of 5 mg/50 ml (1 ml of 5 mg/ml stock in 50 ml buffer).
3. Lay each membrane on the buffer until it wets through and then submerge it. Place an acetate sheet on top to prevent the top membranes from drying out and place a second plastic box top to act as a lid. Incubate at 68 °C for 30 min or longer.

4. Add approximately 10^6 dpm labelled lambda phage DNA to the probe for hybridization to the molecular weight marker lane. Pierce the top of the microfuge tube containing the probe with a needle to vent the tube while heating. Heat the tube at 100 °C in a heat block for 5 to 10 min to denature the probe. Immediately add the probe to the hybridization box or store on ice to slow reannealing of the probe.

5. Remove the acetate sheet and membranes from the hybridization box and set them aside. Add the probe to the hybridization buffer and mix well. Add the membranes back to the box one at a time and mix well in between each addition. Avoid bubbles. Replace the acetate sheet and lid and incubate overnight without shaking at 68 °C.

6. Prepare 1 liter of the first wash buffer (2 × SSC, 0.1% SDS) and heat to 68 °C. Remove the acetate sheet to a container of water for cleaning (the sheets are easily cleaned and can be reused). Pour some of the wash buffer into another hybridization box. Place each membrane followed by a washing screen into the box. Pour off this first wash solution and replace with additional wash solution. Shake for 20 min at 68 °C.

7. Pour off the first wash solution and replace with 1 × SSC, 0.1% SDS at 68 °C. Shake for 20 min at 68 °C.

8. Pour off the second wash solution and replace with 0.5 × SSC, 0.1% SDS at 68 °C. Shake for 20 min at 68 °C.

9. Lay wet membrane on a piece of plastic wrap with the DNA side down. Blot off excess liquid and smooth out any bubbles in between blot and plastic wrap. Wrap blot and place against X-ray film and an intensifier screen in a cassette and store at −80 °C overnight. Longer exposures may be required.

10. After autoradiography is complete, the membranes can be stripped of probe for reuse. Heat a solution of 0.1 × SSC, 0.1% SDS to 80 °C. Wash the filters at this temperature for 10 min. Repeat the wash. Air-dry membranes on filter paper and store.

Source of probes

The most commonly used probes are cloned sequences derived from either mRNA or genomic DNA. The mRNA clones are made by enzymatically converting isolated mRNA into complementary DNA and are referred to as cDNA. cDNA clones vary in length depending on the size of the gene and how completely the mRNA is copied into DNA. Mature mRNA does not contain non-coding intervening sequences (introns) that may interrupt the coding portions of the gene, but they may contain short 5′ and 3′ noncoding sequences. Genomic clones are isolated from chromosomal DNA and contain all sequences, coding and noncoding and may represent random fragments of DNA or be gene-related segments.

There are advantages and disadvantages to using cDNA or genomic clones. Since cDNA clones always represent coding sequences, they tend to be conserved between species and a given cDNA clone may be used on different taxonomic groups. Genomic clones, particularly random clones, may contain a higher proportion of quickly evolving sequences and may not hybridize well to DNA from more distantly related plants. However, if the objective is to work within a single species or a few very closely related species, then genomic clones are often used since they are much simpler to produce than cDNA clones.

The complexity of restriction patterns is an important consideration for simple genetic mapping. A cDNA library may contain many clones that hybridize to single loci and single-copy sequences. The identity of the locus is then confirmed when the probes are applied to genetically diverse plants. For example, if a given probe hybridizes to more than one locus, it may be difficult to determine which bands correspond to which loci when working with different plant materials. Some cDNA clones will represent genes that exist as members of multigene families, any member of which may potentially hybridize to all other members. As stated previously, multiple locus hybridization, whether from cDNA or genomic probes, can sometimes be reduced by hybridizing or washing at high stringencies. It is also possible that unique genomic sequences lying in the vicinity of members of multigene families can be used as locus-specific probes for that family member.

Plant genomes often contain large amounts of dispersed repeated sequences and genomic clones containing such sequences give very complex restriction patterns. It may be possible to reduce the signal from repeated elements that form part of a probe by blocking the membrane with homologous repeat DNA during hybridization and prehybridization. However, if truly single or low-number locus probes are desired, then a random genomic library needs to be screened to eliminate these complex probes. Often such libraries are constructed from 0.5 to 2.0 kb DNA fragments in order to obtain sequences that are free of interspersed repeated DNA. These libraries are usually screened against labelled total plant DNA with the result that those clones that give appreciable hybridization signal probably contain repeated sequences and are discarded. It has been reported that genomic libraries constructed from methylation-sensitive enzymes, such as *Pst* I, are enriched for undermethylated DNA and have a high proportion of single or low copy sequences [8]. This has been demonstrated in tomato where a *Pst* I-generated library gave more single-copy clones than one generated with *Eco* RI [10]. However, both enzymes were equivalent in producing low-copy probes in lentil [7].

In maize, genomic clones often give better hybridization signals than cDNA clones because they are contiguous with the target sequence whereas the presence of introns in a gene will interfere with the hybridization of cDNA clones [8]. Genomic clones are also often longer than cDNA clones and the greater length results in more probe being bound to restriction fragments and a greater hybridization signal.

It is desirable to have probes that detect high levels of polymorphism. In this regard, cDNA clones may be better than random genomic clones. In lettuce [9], lentil [7], and tomato [10] cDNA probes were more efficient in detecting polymorphisms than genomic probes but in potato this difference was not observed [5].

Solutions

- Gel loading buffer
  - 7.0 ml glycerol
  - 0.5 ml 10X neutral electrophoresis buffer
  - 0.1 ml 20% SDS (sodium dodecyl sulfate)
  - 0.8 ml 0.25 M EDTA
  - 0.2 ml 10 mg/ml bromophenol blue (tracking dye)
  - 1.4 ml H_2O
  - Final volume 10 ml
- 10X neutral electrophoresis buffer (NEB), 10 l
  - 1210.0 g Tris (Sigma 7-9)
  - 33.6 g EDTA, disodium salt
  - 170.1 g sodium acetate trihydrate

Dissolve components in H_2O to an approximate volume of 9.5 l. Adjust pH to 8.1 with glacial acetic acid and bring final volume to 10 l.
- Ethidium bromide
  - 1.0 g ethidium bromide
  - 100 ml H_2O

Stir many hours to dissolve. Store in a light-proof bottle (i.e. wrap in aluminum foil). Ethidium bromide is a mutagen and gloves should be worn when handling it. Do not breath dust.
- 10X acid blotting solution
  - 2.5 M HCl
- 5X base blotting solution
  - 2.0 M NaOH
- 20X SSC
  - 3.0 M NaCl
  - 0.3 M sodium citrate
  - pH to 7.0
- 10X random primers labelling buffer
  - 900 mM HEPES pH 6.6
  - 100 mM $MgCl_2$
  - 20 mM dithiothreitol

Store frozen in small aliquots
- dATP, dGTP, TTP

Make individual 15 mM stocks of these deoxynucleotide triphosphates by dissolving in sterile H_2O and bringing to a neutral pH with 150 mM Trizma base. Monitor the pH by spotting small amounts on pH indicator paper. Combine equal amounts of all three triphosphates and store frozen in 100 μl aliquots for use with labelling with radioactive dCTP.
- Sephadex G50-80

Swell 5 g Sephadex G50-80 by placing it in 100 ml 1% SDS, 25 mM EDTA (column buffer) overnight.
- Column dye

Dissolve 20 mg blue dextran in 1 ml column buffer (see above). Add 25 μl of 10 mg/ml stock of bromophenol blue.
- Hybridization buffer
  Add in order:
  - 490 ml distilled water
  - 250 ml 20X SSC
  - 50 ml 1.0 M sodium phosphate pH 7.2
  - 10 ml 0.25 M EDTA
  - 50 ml 100× Denhardt's solution
  - 30 ml 20% SDS (w/v)
  - 100 ml 50% (w/v) high-molecular-weight dextran sulfate
  - Store in a refrigerator
- 100× Denhardt's solution
  - 10 g Ficoll (Sigma, type 400-DL)
  - 10 g polyvinylpyrrolidone (Sigma, PVP-40)
  - 10 g bovine serum albumin (Fraction V)
  - 500 ml H_2O
  The solution should be stored frozen in 50 ml aliquots
- Salmon sperm DNA
  Dissolve DNA in H_2O at a rate of 5 mg/ml. It may take many hours with stirring to dissolve. Shear the DNA by sonication to 500–1000 bp. Denature the DNA in 10 ml aliquots by boiling for 10 min. Cool quickly on ice and store frozen.

References

1. Anderson MML, Young BD (1985) Quantitative filter hybridization. In: Hames BD, Higgins SJ (eds) Nucleic Acid Hybridization: A Practical Approach, pp. 73–112. Oxford: IRL Press.
2. Bernatzky R (1988) Restriction fragment length polymorphisms. In: Gelvin SB, Schilperoort RA, Verma DPS (eds) Plant Molecular Biology Manual, pp. C2/1–18. Martinus Nijhoff, Dordrecht, Netherlands.
3. Bernatzky R, Tanksley SD (1986) Methods for detecting single or low copy sequences in tomato on Southern blots. Plant Mol Biol Rep 4: 37–41.
4. Feinberg AP, Vogelstein B (1984) A technique for radiolabelling DNA restriction fragments to high specific activity. Anal Biochem 132: 6–13.
5. Gebhardt C, Ritter E, Debener T, Schachtschabel U, Walkemeier B, Uhrig H, Salamini F (1989) RFLP analysis and linkage mapping in *Solanum tuberosum*. Theor Appl Genet 78: 65–75.
6. Hames BD, Higgins SJ, eds (1985) Nucleic Acid Hybridization: A Practical Approach. Oxford: IRL Press.
7. Havey MJ, Muehlbauer FJ (1989) Linkages between restriction fragment length, isozyme, and morphological markers in lentil. Theor Appl Genet 77: 395–401.
8. Helentjaris T (1987) A genetic linkage map in maize based on RFLPs. Trends Genet 3: 217–221.
9. Landry BS, Kesseli R, Leung H, Michelmore RW (1987) Comparison of restriction endonucleases and sources of probes for their efficiency in detecting restriction fragment length polymorphisms in lettuce (*Lactuca sativa* L.). Theor Appl Genet 74: 646–653.
10. Miller JC, Tanksley SD (1990) Effect of different restriction enzymes, probe source, and probe length on detecting restriction fragment length polymorphism in tomato. Theor Appl Genet 80: 385–389.
11. Pichersky E, Hoffman NE, Malik VS, Bernatzky R, Tanksley SD, Szabo L, Cashmore AR (1987) The tomato *Cab*-4 and *Cab*-5 genes encode a second type of CAB polypeptides localized in Photosystem II. Plant Mol Biol 9: 109–120.
12. Reed KC, Mann DA (1985) Rapid transfer of DNA from agarose gels to nylon membranes. Nucleic Acids Res 13: 7207–7221.
13. Rigby PWJ, Dieckman M, Rhodes C, Berg P (1977) Labelling deoxyribonucleic acid to high specific activity *in vitro* by nick translation with DNA polymerase I. J Mol Biol 113: 815–817.
14. Sambrook J, Fritsch EF, Maniatis T (1989) Molecular Cloning: A Laboratory Manual, 2nd ed. Cold Spring Harbor, NY: Cold Spring Harbor Press.
15. Southern EM (1975) Detection of specific sequences among DNA fragments separated by gel electrophoresis. J Mol Biol 98: 503–517.
16. Tautz D, Renz M (1983) An optimized freeze-squeeze method for the recovery of DNA from agarose gels. Anal Biochem 132: 14–19.
17. Wahl GM, Stern M, Stark GR (1979) Efficient transfer of large DNA fragments from agarose gels to diazobenzyloxymethyl-paper and rapid hybridization using dextran sulfate. Proc Nat Acad Sci USA 76: 3683–3687.
18. Wetmur JG (1975) Acceleration of DNA renaturation rates. Biopolymers 14: 2517.
19. Williams PH, Hill CB (1986) Rapid cycling populations of *Brassica*. Science 232: 1385–1389.

3. Organelle DNA isolation and RFLP analysis

JEFFREY D. PALMER

Department of Biology, Indiana University, Bloomington, IN 47405, USA

Introduction

The analysis of restriction fragment length polymorphisms (RFLPs) in organelle DNA is radically different from that in nuclear DNA in both purpose and
execution. The fundamental reason for these differences is genome size: in land
plants, chloroplast DNA (cpDNA) is $1-2 \times 10^2$ kb in size [38, 39] and mitochondrial DNA (mtDNA) $2-20 \times 10^2$ kb [14, 26, 37, 39], whereas most
nuclear genomes are $1-10 \times 10^6$ kb [2]. The larger size of nuclear genomes is
reflected in our relatively poor understanding of their structure and the fact that
only a small fraction of nuclear genes have been isolated and analyzed. Physical
mapping of RFLPs for an entire, average-sized nuclear genome is a forbidding
task whose completion is at least several years away, and the physical mapping
of even a 1% region of the genome presents a significant challenge. As a
consequence, nuclear RFLPs are presently used primarily as genetic markers
in linkage studies aimed at identifying and selecting traits of agronomic importance [27, 44] and at examining modes of chromosomal evolution [15, 54].
Nuclear RFLPs are just beginning to be used as physical markers in chromosome walking efforts to clone mapped genes of interest [58].

By contrast, the small size of organelle DNAs has allowed the construction
of complete physical maps for hundreds of cpDNAs and dozens of mtDNAs
[27, 37–39]. Moreover, three entire chloroplast genomes [17, 31, 50] and many
mitochondrial genes [14, 26] have been sequenced. Consequently, genetic mapping of RFLPs to identify and isolate organelle genes is simply unnecessary,
not to mention impossible in land plants given the clonal inheritance of
organelle genomes in sexual crosses [13, 49]. Instead, RFLPs are used principally in physical studies aimed at understanding the structure and evolution
of organelle genomes [33, 34, 38, 39], their phylogeny [6, 36, 43, 52], and their
inheritance [13, 30]. Organelle RFLPs have been used as genetic markers in
but two special circumstances, in studies of cpDNA recombination in sexual
crosses of the green algal genus *Chlamydomonas* [13, 25] and in investigations
of mtDNA recombination in somatic hybrids of angiosperms [46, 47].

The execution of RFLP studies is considerably facilitated by the small size
of organelle genomes and their relative ease of enrichment and purification. The
first part of this chapter presents several procedures for purifying plant
organelle DNAs and offers a variety of modifications and alternative
approaches. The merits of using total cellular DNA instead of purified organelle

35

J.S. Beckmann and T.C. Osborn (eds) Plant Genomes: Methods for Genetic and
Physical Mapping, 35–53.
© 1992 *Kluwer Academic Publishers.*

DNA for RFLP studies are also considered. In the second part of the chapter, I briefly discuss approaches for analyzing organelle RFLPs and the uses to which they have been applied.

Procedures

Isolation of organelle DNA

In general, strategies for purifying organelle DNAs differ markedly between land plants and algae. In land plants, chloroplasts and mitochondria are often readily isolated in reasonably intact form, permitting the concomitant enrichment of their constituent genomes, which otherwise are not easily separable from nuclear DNA in dye-CsCl gradients. In algae, however, the organelles are often much harder to isolate, whereas their genomes are often readily purified from preparations of total cellular DNA by banding in dye-CsCl gradients.

Below I describe two procedures for isolating chloroplasts from land plants. The first involves the physical separation of chloroplasts from nuclear and other cellular material by sucrose step gradient centrifugation [32, 35, 55]. We find this to be the most generally applicable procedure for preparing DNA that is substantially enriched, although usually not completely pure, in cpDNA sequences. This method works well for many land plants and a few algae with large, easily fractured cells (e.g. *Nitella*). The second procedure involves the enzymatic treatment of a crude chloroplast preparation with DNAse I to obtain cpDNA of generally quite high purity [22, 23]. This procedure has the major drawback of giving very low yields of DNA and is reserved for those situations where large amounts of plant material are available. Protocols are also presented for purifying mtDNA from land plants (by DNAse I treatment of the organelles [21]) and for purifying both organelle DNAs from algae (by fractionation in dye-CsCl gradients [8, 9]).

It is important to recognize that just because it is theoretically, and often practically, possible to purify organelle DNA, in many cases this is not sufficient reason to actually do so. Because the circular chromosomes of chloroplasts and mitochondria are present in multiple copies in every leaf cell (on average several thousands of copies for the former and a few hundred for the latter [1]), their sequences are easily visualized in preparations of total cellular DNA by the Southern hybridization approach that is at the heart of RFLP analysis. In fact, using total DNA rather than enriched or purified organelle DNA provides several major advantages. These include considerations of *yield* (total DNA extraction efficiencies are generally at least 5–10 times higher than for organelle DNA extraction; 1–2 g of fresh leaf material will usually provide sufficient total DNA for an organelle DNA RFLP study), *flexibility* (total DNA can be used to study all three plant genomes), and *adaptability* (in several cases total DNA has been successfully extracted from plants for which organelle DNA extraction was unsuccessful). Rapid, small-scale procedures for preparing total DNA from plants are described by Kidwell and Osborn in this book and by others elsewhere [12, 43].

Protocol 1: Isolation of cpDNA using sucrose step and CsCl-ethidium bromide gradients

Steps in the procedure

1. Use young, unexpanded green leaves if at all possible since they will have smaller cells than older fully expanded leaves and will yield more DNA. Occasionally it is useful to place plants in the dark for 1–4 days prior to extraction to reduce chloroplast starch levels; however, this is usually either unnecessary or impractical.
2. Cut leaves into small pieces (2–10 cm² in surface area). Wash cut leaves in tap water only if visibly dirty or buggy.
3. Place 10–100 g of cut leaves in 50–400 ml of ice-cold isolation buffer (Appendix).
4. Homogenize in blender for three to five 5-second bursts at high speed. A longer homogenization period is occasionally necessary.
5. Filter through four layers of cheesecloth (with squeezing).
6. Centrifuge filtrate at 1000 g for 15 min at 4 °C.
7. Resuspend the pellet from 10–50 g of starting material in 5–8 ml of ice-cold wash buffer (Appendix) using a soft paint brush and vigorous swirling.
8. Load the resuspended pellet onto a step gradient consisting of 17 ml of 52% sucrose overlayed with 8 ml of 30% sucrose, both in 50 mM Tris-HCl, pH 8.0, 25 mM EDTA. The overlay should be added with sufficient mixing to create a diffuse interface and thereby prevent trapping of nuclear material in the band of chloroplasts that forms at the 30–52% interface. Best results are obtained when the step gradient is poured the day prior to extraction.
9. Centrifuge the step gradients at 25,000 rpm for 30–60 min at 4 °C in a SW-27 (Beckman) or AH-627 (Sorvall) rotor.
10. Remove chloroplast band from the 30–52% interface using a wide-bore pipette, dilute with 3–10 volumes wash buffer, and spin at 1,500 g for 15 min at 4 °C.
11. Resuspend chloroplast pellet in 1–2 ml wash buffer (or 15 ml for TV-850 gradient).
12. Add 1/20 volume of a 20 mg/ml solution of self-digested (2 h at 37 °C) proteinase and incubate for 2–10 min at room temperature.
13. Gently add one-fifth volume of lysis buffer (see Solutions). Slowly invert the tube several times over a period of 15–30 min at room temperature.
14. A cpDNA-enriched 'total' DNA preparation can be prepared by resuspending the pellet of the sucrose gradient in 1.5 ml wash buffer, lysing (steps 12 and 13), clearing (centrifuge at 1,000 g for 10 min to pellet starch and cellular debris), and CsCl-banding (see below).
15. Bring the chloroplast DNA sample (or mitochondrial or whole cell lysate, see below) to a volume of roughly 3 ml. Add 3.35 g technical grade CsCl (Kawecki Berylco Industries Inc.), freshly powdered in a mortar and pestle, and dissolve by gentle mixing. Add ethidium bromide to a final concentration of 200 μg/ml and distilled H_2O to bring to a final volume of 4.45 ml and a final density of 1.55 g/ml.

16. Centrifuge in a TV-865 or TV-1665 rotor (Sorvall) for 4–16 h at 50,000–58,000 rpm at 20 °C. We prefer the Sorvall rotor over the equivalent Beckman one because of the simpler tube capping system and because the centrifuge tubes can be reused as many as 10–15 times. However, centrifuge tubes should not be reused if the DNA is to be used for PCR studies.

17. Remove any scum (this will be considerable in the case of a directly banded chloroplast lysate) from the top of the gradient using a 1-ml pipette tip with the end cut off. Use a second pipette tip with the end cut obliquely to remove the visible band of DNA. This should be removed in as small a volume as possible, 0.5–1.0 ml.

18. If the DNA fraction is visibly dirty after the first gradient (as is often the case with direct banding of chloroplast lysates), it can be banded a second time. Bring the DNA/CsCl fraction to a volume of 4.45 ml by adding a premixed solution of CsCl (density of 1.55 mg/ml = 750 mg powdered CsCl added per ml final volume), ethidium bromide (100 μg/ml), and TE and repeat steps 16 and 17.

19. Remove ethidium bromide by three extractions with isopropanol (uppermost layer) saturated with NaCl (bottom) and H_2O (middle). For each extraction add 1–2 ml isopropanol, mix gently by inversion (shake moderately if necessary), and let stand for phase separation. Then remove the isopropanol. After the last extraction spin for 2 min in a table-top centrifuge at 200–500 *g* to ensure good phase separation.

20. There are two ways to remove the CsCl. Either dialyze against at least three changes of 2 liters of TE over a period of 1–2 days or ethanol-precipitate as described in steps 21–24.

21. Remove the aqueous layer from the third isopropanol extraction and add two volumes of H_2O to dilute the CsCl. Mix gently and add 6 volumes of ice-cold ethanol to precipitate DNA. Place at −20 °C for 30 min to overnight. Do not place at −80 °C or the CsCl will precipitate.

22. Spin at 2,000 *g* or greater for 10 min to collect the DNA precipitate.

23. Wash pellet with 76% ethanol. Spin at 2,000 *g* for 2 min to collect the DNA.

24. Resuspend pellet in 0.1–0.5 ml TE.

25. Store the DNA at 4 °C for short-term use and at −20 °C for long-term use.

Protocol 2: DNAse I isolation of cpDNA

Steps in the procedure
1. Place 0.1–1.0 kg of chopped green leaves (see Protocol 1, steps 1 and 2) in a 3–10-fold excess (w/v) of ice-cold isolation buffer.
2. Homogenize, filter, and centrifuge as in steps 4–6 of Protocol 1.
3. Resuspend the pellet in 200 ml of DNAse I buffer (see Solutions) per kg of starting material using a soft paint brush and vigorous swirling.
4. Add 15 mg of DNAse I (freshly dissolved in DNAse I buffer) per kg of starting material to the resuspended chloroplasts, mix gently but well, and incubate for 1 h on ice with occasional mixing.
5. Add three volumes of wash buffer and centrifuge at 1,500 *g* for 15 min at 4 °C.
6. Resuspend pellet in wash buffer and repeat washing spin twice more.
7. Resuspend pellet, lyse chloroplasts, and purify cpDNA by CsCl centrifugation as described in steps 11–25 of Protocol 1.

Protocol 3: DNAse I isolation of mtDNA

Steps in the procedure
1. Place 0.1–1.0 kg of chopped green leaves or etiolated shoots (see Protocol 1, steps 1 and 2) in a 3–10-fold excess (w/v) of ice-cold isolation buffer.
2. Homogenize, filter, and centrifuge as in steps 4–6 of Protocol 1.
3. Discard pellet (or use to prepare chloroplasts; see Protocols 1 or 2), and, if necessary, repeat low-speed spin to remove residual chloroplast and nuclear material.
4. Pellet mitochondria by centrifugation at 12,000 *g* for 20 min at 4 °C.
5. Resuspend the pellet in 100 ml of DNAse I buffer per kg of starting material using a soft paint brush and vigorous swirling.
6. Add 30 mg DNAse I (freshly resuspended in DNAse I buffer) per kg of starting material, mix gently but well, and incubate for 1 h on ice with occasional mixing.
7. Add three volumes wash buffer and centrifuge at 12,000 *g* for 20 min at 4 °C.
8. Resuspend pellet in wash buffer and repeat washing spin twice more.
9. Resuspend pellet, lyse mitochondria, and purify mtDNA as described in steps 11–25 of Protocol 1 for cpDNA.

Protocol 4: Dye-gradient purification of algal organelle DNAs

Steps in the procedure

1. Harvested algal cells or blades are homogenized in various ways (from gentle swirling in detergent-containing lysis buffer, to grinding with a mortar and pestle, to mechanical homogenization with a Dounce homogenizer, tissue homogenizer or French pressure cell) according to the type of tissue (for specific examples, see [8, 9]).
2. Treat homogenate with proteinase and lyse as in steps 12 and 13 of Protocol 1.
3. Add CsCl and ethidium bromide and purify total cell DNA via density gradient centrifugation as in steps 15—19 of Protocol 1. Be sure to take a broad region of the gradient containing the main-band DNA to ensure recovery of organelle DNAs.
4. After removal of ethidium bromide, add CsCl to 1.66 g/ml (refractive index = 1.3960) and Hoechst 33258 dye (bisbenzimide) to 100 μg/ml, and spin in TV-865 rotor for 4—16 h at 50,000—58,000 rpm at 20 °C.
5. Remove the various visible DNA bands (either directly, as described in step 17 of Protocol 1, or by drip-fractionation of the gradient). Algal cpDNAs and mtDNAs are almost always substantially more AT-rich than nuclear DNAs and will therefore bind more Hoechst dye and band higher in the gradient, forming minor upper bands relative to the major lower-band nuclear DNA. The relative positions of the cpDNA and mtDNA bands (as well as other minor bands resulting from bacterial contamination or nuclear rDNA satellites) vary from species to species [8, 9].
6. If necessary, either to achieve further purification of the various DNA fractions or simply to combine and concentrate them, pool all of the same fractions collected from different initial gradients and centrifuge again as in step 4.
7. Remove Hoechst dye and CsCl as in steps 19—24 of Protocol 1.

Modifications and alternative procedures

An earlier account [32] of the sucrose step gradient procedure for chloroplast isolation (Protocol 1) included detailed comments on factors that can influence its success. These factors include 1) the importance of starting with the freshest possible leaf material; 2) the addition of various components to the isolation buffer (PEG, PVP, DTT, antifoaming agents); 3) the length and force of homogenization and ratio of tissue to buffer during homogenization; and 4) the exact composition of the sucrose gradient and amount of material loaded per gradient. The homogenization step can be particularly crucial; in some cases we have found that much higher yields are obtained when fresh leaf material is frozen in liquid nitrogen, powdered (using a mortar and pestle or a mill [5]), and then hydrated in isolation buffer by homogenization in a blendor or tissue homogenizer. Adaptations of the basic sucrose step gradient procedure designed specifically for certain conifers [56] and mosses [5] have also been described.

Mourad and Polacco [29] recently reported that substitution of nuclease-free proteinase K for the more commonly used pronase at the step immediately prior to lysis gave enhanced yields of cpDNA from DNAse I-treated chloroplasts. On the other hand, Hsu and Mullin [18] obtained cleaner mtDNA that was less contaminated with nuclear DNA, less colored, and more digestible with restriction enzymes, when proteinase K was omitted at the lysis stage. These investigators also reported the results of varying several other parameters, including ionic strength of the homogenization buffer, extent and force of homogenization, and the detergent composition of the lysis buffer. To maximize yields from very small amounts of starting material, Kemble [20] reported modifications of both the sucrose gradient (for cpDNA) and DNAse I (for mtDNA) procedures.

Palmer [35] reviewed several alternative gradient techniques, most of which are seldom used anymore, to the sucrose step gradient procedure described in Protocol 1. The use of Percoll gradients [16, 45] for preparing chloroplasts is, however, commonplace in experiments that require highly intact and active chloroplasts. RFLP studies of large numbers of taxa generally do not warrant the extra expense (of the Percoll gradient medium) and effort (to determine the proper gradient configuration for each plant) required. Bowman and Dyer [4] introduced to cpDNA RFLP analysis the use of a nonaqueous gradient technique for banding chloroplasts from a homogenate of freeze-dried (lyophilized) leaf tissue. Dally and Second [7] reported significant modifications of this approach, including the adoption of substantially less hazardous organic components of the density gradient and the use of Triton X-100 in the lysis medium to differentially solubilize cpDNA and nuclear DNA.

Several approaches to achieving substantial enrichment of cpDNA without resorting to either density gradient banding or DNAse I treatment of chloroplasts are also known. In some cases [42], one can obtain several-fold enrichment of cpDNA sequences very quickly and easily simply by performing Protocol 1 as described, except omitting the entire sucrose step gradient step and (sometimes) the subsequent washing step. Bookjans *et al.* [3] described a similarly rapid no-gradient procedure based on the use of a high-ionic-strength (1.25 M NaCl) isolation medium. This NaCl procedure has been modified in two recent reports to include the use of hexadecyltrimethylammonium bromide (CTAB) as a component of either the lysis [28] or the immediately post-lysis [48] media.

Solutions

— Isolation buffer
  — 0.35 M sorbitol
  — 50 mM Tris-HCl, pH 8.0
  — 5 mM EDTA
  — 0.1% BSA (w/v)
  — 0.1% β-mercaptoethanol (v/v; add fresh to isolation buffer immediately prior to use)
— Wash buffer
  — 0.35 M sorbitol

- – 50 mM Tris-HCl, pH 8.0
- – 25 mM EDTA
- – Lysis buffer
  - – 5% sodium sarcosinate (w/v)
  - – 50 mM Tris-HCl, pH 8.0
  - – 25 mM EDTA
- – DNAse I buffer
  - – 0.35 M sorbitol
  - – 50 mM Tris-HCl, pH 8.0
  - – 15 mM MgCl$_2$
- – TE
  - – 10 mM Tris-HCl, pH 8.0
  - – 1 mM EDTA

RFLP analysis of organelle DNA

Southern blot hybridizations are crucial to almost all RFLP studies. The high copy number of organelle genomes [1] relative to single-copy nuclear genes means that strong hybridization signals are usually obtained, even when quite divergent heterologous probes are used or when hybridization is to total DNA rather than organelle DNA. Bidirectional, 'sandwich' blotting [51] is now routinely performed to obtain two identical filter blots from each gel.

In land plants, cpDNA and mtDNA differ dramatically in their tempo and mode of evolutionary change [14, 37–39, 41, 57]. CpDNA evolves slowly in most respects, including genome size, gene order, and primary sequence. MtDNA changes even more slowly in primary sequence than cpDNA, but is subject to huge and rapid changes in size and sequence arrangement. As described in the following sections, these differences in evolutionary behavior significantly affect the approaches one adopts for analyzing RFLPs, as well as the uses to which such studies are applied.

CpDNA RFLPs

As stated in the Introduction, RFLP analysis of organelle genomes is generally performed for different reasons than for nuclear DNA. Of the several applications for organelle DNA RFLP analysis, the one that is most actively pursued and that involves the most sophisticated analysis is the study of cpDNA evolution, including phylogeny. Because cpDNA is quite invariant in size and arrangement compared to both mtDNA and nuclear DNA, restriction mapping approaches can generally be used to uncover the underlying mutational basis of cpDNA RFLPs. These mutations can then be used to better understand the process of cpDNA evolution and as characters in the reconstruction of phylogeny. In contrast, it is usually difficult, if not impossible, to accurately assess the mutational basis of RFLPs in mtDNA and nuclear DNA, in which case they can serve only as blind, although clearly quite useful in the case of nuclear DNA, markers of genetic diversity.

Most cpDNA mutations, including the majority of those detected by RFLP analysis, are point mutations (apparent as the loss/gain of restriction sites). A less common, but still frequent class of mutations is small-length mutations (deletions/insertions) of 1–1,000 bp in size [33, 34, 38]. Most of these are too small (less than 50 bp in length) to be detected with the 6 bp restriction enzymes typically used in cpDNA RFLP studies. However, even the residual larger-length mutations are still commonly encountered in comparing cpDNAs of congeneric species and even, sometimes, conspecific populations. A third, exceedingly rare class of mutations is inversions [33, 34, 38], which are recognizable as the movement (inversion) of restriction sites unaccompanied by any net gain or loss in size or number of sites in the affected region. Only a single inversion has yet been detected in standard cpDNA RFLP studies [42], however these mutations can be useful as phylogenetic markers at deeper evolutionary levels, as described in the last paragraph of this section.

The predominant use of cpDNA RFLP analysis – to reconstruct plant phylogeny – has been reviewed elsewhere with respect to experimental strategy, data analysis, applications, and the kinds of information obtained [6, 10, 36, 43, 52]. I will therefore provide only a brief summary of the approach. The general aim is to use restriction mapping procedures to detect sufficient numbers of phylogenetically informative restriction site mutations to enable construction of a highly resolved phylogeny. Length mutations are often excluded from formal phylogenetic analysis because of the high levels of homoplasy associated with them [42, 43]. It is nonetheless important to properly diagnose length mutations as such, so as to avoid erroneous overcounting of them as multiple site changes. Generally a dozen or so 6 bp cutting restriction enzymes are used in these RFLP mapping studies, each enzyme cutting 20–100 times in the chloroplast genome. Usually, a set of 4–40 probes covering the entire chloroplast genome is used in sequential hybridizations to the set of two filters (per enzyme blot) generated by double-sided blotting. Fewer, larger probes can be used in situations where little variation is expected, whereas smaller ones are necessary to resolve more complex situations of variation. Clone banks are now available from a large variety of land plant cpDNAs (see [35, 43] for listings); most are suitable across a wide range of taxa owing to the conservative nature of cpDNA sequence and structural evolution. Perhaps the most generally useful library is one available from this author, which consists of 40 abutting clones, of average size 3–4 kb, which together span the entire tobacco chloroplast genome. The tobacco genome is a useful reference because it is one of the three completely sequenced cpDNAs [50] and because it has the same gene order as most other vascular plant cpDNAs [37–39]. The construction of restriction site maps from these hybridization experiments and the use of restriction site mutations as phylogenetic characters are discussed elsewhere [10, 43].

'Major' chloroplast gene rearrangements are defined as 1) inversions, 2) deletions/insertions of introns, 3) partial or complete deletions/insertions of genes, and 4) deletion of one segment of the large inverted repeat found in most chloroplast genomes. These mutations are quite rare relative to point mutations and therefore cannot be expected to produce highly resolved phylogenies. However, when found, they hold great promise for illuminating specific and deep branchings of plant evolution. Examples of RFLP surveys to detect and circumscribe individual gene rearrangements [19, 24] and general approaches for surveying for them and lists of them [10, 11, 38, 43] have been published.

MtDNA RFLPs

The high rate of rearrangement and size change in plant mtDNA makes the molecule decidedly less useful than cpDNA for phylogenetic RFLP studies. The rampant structural variation in mtDNA makes it difficult, if not impossible, to align homologous sequences for comparison of restriction sites. Although large inversions can themselves serve as useful phylogenetic characters, as described in the preceding paragraph, the high frequency with which they occur in plant mtDNA severely limits their utility. Later inversions tend to obscure earlier ones and the amount of work required for proper analysis is often not rewarded at those taxonomic levels at which they occur with sufficient infrequency. Small discrete rearrangements, such as the loss/gain of introns and genes, appear to be less common in mtDNA [40] than in cpDNA and therefore are likely to provide only very limited amounts of phylogenetic information.

From a purely technical side, the analysis of mtDNA RFLPs has two limitations compared to cpDNA. First, a major caveat to the use of total DNA to analyze mtDNA RFLPs is that land plant mtDNAs are notorious for containing numerous pieces of stably integrated, largely functionless and duplicate, cpDNA sequences [14, 26, 39, 53]. Since cpDNA is usually present in 10–50 times higher copy number than mtDNA in leaves [1], the use of mtDNA fragments containing such cpDNA integrants as hybridization probes can unwittingly reveal cpDNA polymorphism in addition to, or even instead of, mtDNA polymorphism. Second, although a number of mtDNAs have been more or less entirely cloned, these libraries are generally useful over only a narrow phylogenetic range. This is because genome size is so variable that many fragments from one genome simply lack significant homologues in another, and because rampant rearrangement will often cause a 5–20 kb fragment from one genome to hybridize to a multitude of different fragments and regions from a second. These considerations suggest that the best probes for mtDNA RFLP studies will be small fragments specific for the 20 or so mitochondrial protein genes and 20–30 rRNA and tRNA genes [26].

Overall, RFLP analysis of plant mtDNA seems destined for a few specialized applications, primarily involving the use of inversion-based RFLPs as blind markers for patterns of mtDNA inheritance in sexual crosses [30], mtDNA recombination in somatic hybrids [46, 47], and, perhaps, populational and speciational divergence in certain situations where cpDNA variation is too limited to provide sufficient phylogenetic resolution.

Acknowledgements

I am grateful to Stephen Downie for critical reading of the manuscript. This work was supported in part by grants from the National Science Foundation (BSR-89–96262) and the National Institutes of Health (GM-35087).

Note added in proof
This article was written in the first half of 1990.

References

1. Bendich AJ (1987) Why do chloroplasts and mitochondria contain so many copies of their genome? Bioessays 6: 279–282.
2. Bennett MD, Smith JB (1976) Nuclear DNA amounts in angiosperms. Phil Trans Roy Soc Ser B 274: 227–274.
3. Bookjans G, Stummann BM, Henningsen KW (1984) Preparation of chloroplast DNA from pea plastids isolated in a medium of high ionic strength. Anal Biochem 141: 244–247.
4. Bowman DM, Dyer TA (1982) Purification and analysis of DNA from wheat chloroplasts isolated in nonaqueous media. Anal Biochem 122: 108–118.
5. Calie PJ, Hughes KW (1987) An efficient protocol for the isolation and purification of chloroplast DNA from moss gametophyte tissues. Plant Mol Biol Rep 4: 206–212.
6. Crawford DJ (1990) Plant Molecular Systematics: Macromolecular Approaches. New York: Wiley.
7. Dally AM, Second G (1989) Chloroplast DNA isolation from higher plants: An improved non-aqueous method. Plant Mol Biol Rep 7: 135–143.
8. Delaney TP, Cattolico RA (1989) Chloroplast ribosomal DNA organization in the chromophytic alga *Olisthodiscus luteus*. Curr Genet 15: 221–229.
9. Douglas SE (1988) Physical mapping of the plastid genome from the chlorophyll c-containing alga, *Cryptomonas* sp. Curr Genet 14: 591–598.
10. Dowling TE, Moritz C, Palmer JD (1990) Nucleic acids II: restriction site analysis. In: Hillis D, Moritz C (eds) Molecular Systematics, pp. 250–317. Sunderland: Sinauer.
11. Downie SR, Palmer JD (1992) Use of chloroplast DNA rearrangements in reconstructing plant phylogeny. In: Soltis PS, Soltis DE, Doyle JJ (eds) Molecular Systematics of Plants, pp. 14–35. New York: Chapman and Hall.
12. Doyle JJ, Doyle JL (1987) A rapid DNA isolation procedure for small quantities of fresh leaf tissue. Phytochem Bull 19: 11–15.
13. Gillham NW, Boynton JE, Harris EH (1991) Transmission of plastid genes. In: Bogorad L, Vasil IK (eds.) Cell Culture and Somatic Cell Genetics of Plants, Vol. 7A, The Molecular Biology of Plastids, pp. 55–92. New York: Academic Press.
14. Gray MW (1989) Origin and evolution of mitochondrial DNA. Annu Rev Cell Biol 5: 25–50.
15. Helentjaris T, Weber D, Wright S (1988) Identification of the genomic locations of duplicate nucleotide sequences in maize by analysis of restriction fragment length polymorphisms. Genetics 118: 353–363.
16. Herrmann RG (1982) The preparation of circular DNA from plastids. In: Edelman M, Hallick R, Chua NH (eds) Methods in Chloroplast Molecular Biology, pp. 259–280. Amsterdam: Elsevier.
17. Hiratsu J, Shimada H, Whittier RF, Ishibashi T, Sakamoto M, Mori M, Kondo C, Honji Y, Sun CR, Meng BY, Li YQ, Kanno A, Nishizawa Y, Hirai A, Shinozaki K, Sugiura M (1989) The complete sequence of the rice (*Oryza sativa*) chloroplast genome: Intermolecular recombination between distinct tRNA genes accounts for a major plastid DNA inversion during the evolution of the cereals. Mol Gen Genet 217: 185–194
18. Hsu CL, Mullin BC (1988) A new protocol for isolation of mitochondrial DNA from cotton seedlings. Plant Cell Rep 7: 356–360.
19. Jansen RK, Palmer JD (1987) A chloroplast DNA inversion marks an ancient evolutionary split in the sunflower family (Asteraceae). Proc Natl Acad Sci USA 84: 5818–5822.
20. Kemble RJ (1987) A rapid, single leaf, nucleic acid assay for determining the cytoplasmic organelle complement of rapeseed and related *Brassica* species. Theor Appl Genet 73: 364–370.
21. Kolodner R, Tewari KK (1972) Physicochemical characterization of mitochondrial DNA from pea leaves. Proc Natl Acad Sci USA 69: 1830–1834.
22. Kolodner R, Tewari KK (1975) The molecular size and conformation of the chloroplast DNA from higher plants. Biochim Biophys Acta 402: 372–390.
23. Kolodner R, Tewari KK, Warner RC (1976) Physical studies on the size and structure of the

covalently closed circular chloroplast DNA from higher plants. Biochim Biophys Acta 447: 144–155.

24. Lavin M, Doyle JJ, Palmer JD (1990) Evolutionary significance of the loss of the chloroplast DNA inverted repeat in the Leguminosae subfamily Papilionoideae. Evolution 44: 390–402.

25. Lemieux C, Turmel M, Seligy VL, Lee RW (1984) Chloroplast DNA recombination in interspecific hybrids of *Chlamydomonas*: Linkage between a non-Mendelian locus for streptomycin resistance and restriction fragments coding for 16S rRNA. Proc Natl Acad Sci USA 81: 1164–1168.

26. Lonsdale DM (1989) The plant mitochondrial genome. In: Marcus A (ed), Biochemistry of Plants, Vol. 15, Molecular Biology, pp. 229–295. New York: Academic Press.

27. Martin B, Nienhuis J, King G, Schaefer A (1989) Restriction fragment length polymorphisms associated with water use efficiency in tomato. Science 243: 1725–1728.

28. Milligan BG (1989) Purification of chloroplast DNA using hexadecyltrimethylammonium bromide. Plant Mol Biol Rep 7: 144–149.

29. Mourad G, Polacco ML (1989) Mini-preparation of highly purified chloroplast DNA from maize. Plant Mol Biol Rep 7: 78–84.

30. Neale DB, Sederoff RR (1988) Inheritance and evolution of conifer organelle genomes. In: Hanover JW, Keathley DE (eds) Genetic Manipulation of Woody Plants, pp. 251–264. New York: Plenum.

31. Ohyama K, Fukuzawa H, Kohchi T, Shirai H, Sano H, Sano S, Umesono K, Shiki Y, Takeuchi M, Chang Z, Aota S, Inokuchi H, Ozeki H (1986) Chloroplast gene organization deduced from complete sequence of liverwort *Marchantia polymorpha* chloroplast DNA. Nature 322: 572–574.

32. Palmer JD (1982) Physical and gene mapping of chloroplast DNA from *Atriplex triangularis* and *Cucumis sativa*. Nucleic Acids Res 10: 1593–1605.

33. Palmer JD (1985) Evolution of chloroplast and mitochondrial DNA in plants and algae. In: MacIntyre RJ (ed) Monographs in Evolutionary Biology: Molecular Evolutionary Genetics, pp. 131–240. New York: Plenum.

34. Palmer JD (1985) Comparative organization of chloroplast genomes. Annu Rev Genet 19: 325–354.

35. Palmer JD (1986) Isolation and structural analysis of chloroplast DNA. Meth Enzymol 118: 167–186.

36. Palmer JD (1987) Chloroplast DNA evolution and biosystematic uses of chloroplast DNA variation. Am Natur 130: S6–S29.

37. Palmer JD (1990) Contrasting modes and tempos of genome evolution in plant organelles. Trends Genet 6: 115–120.

38. Palmer JD (1992) Plastid chromosomes: structure and evolution. In: Bogorad L, Vasil IK (eds) Cell Culture and Somatic Cell Genetics of Plants, Vol. 7A, The Molecular Biology of Plastids, pp. 5–53. New York: Academic Press.

39. Palmer JD (1992) Chloroplast and mitochondrial genome evolution in land olants. In: Herrmann RG (ed) Plant Gene Research, Vol. 6, Organelles, in press. New York: Springer-Verlag.

40. Palmer JD (1992) Mitochondrial DNA in plant systematics: applications and limitations. In: Soltis PS, Soltis DE, Doyle JJ (eds) Molecular Systematics of Plants, pp. 36–49. New York: Chapman and Hall.

41. Palmer JD, Herbon LA (1988) Plant mitochondrial DNA evolves rapidly in structure, but slowly in sequence. J Mol Evol 28: 87–97.

42. Palmer JD, Jorgensen RA, Thompson WF (1985) Chloroplast DNA variation and evolution in *Pisum*: Patterns of change and phylogenetic analysis. Genetics 109: 195–213.

43. Palmer JD, Jansen RK, Michaels HJ, Chase MW, Manhart JR (1988) Chloroplast DNA variation and plant phylogeny. Ann Missouri Bot Gard 75: 1180–1206.

44. Paterson AH, Lander ES, Hewitt JD, Peterson S, Lincoln SE, Tanksley SD (1988) Resolution of quantitative traits into Mendelian factors by using a complete linkage map of restriction fragment length polymorphisms. Nature 335: 721–726.

45. Price CA, Cushman JC, Mendiola-Morgenthaler LR, Reardon EM (1987) Isolation of plastids in density gradients of percoll and other silica sols. Meth Enzymol 148: 157–179.
46. Robertson D, Palmer JD, Earle ED, Mutschler MA (1987) Analysis of organelle genomes in a somatic hybrid derived from cytoplasmic male-sterile *Brassica oleracea* and atrazine-resistant *B. campestris*. Theor Appl Genet 74: 303–309.
47. Rothenberg M, Hanson MR (1987) Recombination between parental mitochondrial DNA following protoplast fusion can occur in a region which normally does not undergo intra-genomic recombination in parental plants. Curr Genet 12: 235–240.
48. Sandbrink JM, Vellekoop P, Van Ham R, Van Brederode J (1989) A method for evolutionary studies on RFLP of chloroplast DNA, applicable to a range of plant species. Biochem Syst Ecol 17: 45–49.
49. Sears BB (1980) The elimination of plastids during spermatogenesis and fertilization in the plant kingdom. Plasmid 4: 233–255.
50. Shinozaki K, Ohme M, Tanaka M, Wakasugi T, Hayashida N, Matsubayashi T, Zaita N, Chunwongse J, Obokata J, Yamaguchi-Shinozaki K, Ohto C, Torazawa K, Meng BY, Sugita M, Deno H, Kamogashira T, Yamada K, Kusuda J, Takaiwa F, Kato A, Tohdoh N, Shimada H, Sugiura M (1986) The complete nucleotide sequence of tobacco chloroplast genome: Its gene organization and expression. EMBO J 5: 2043–2049.
51. Smith GE, Summers MD (1980) The bidirectional transfer of DNA and RNA to nitro-cellulose or diazobenzyloxymethyl paper. Anal Biochem 109: 123–129.
52. Soltis PS, Soltis DE, Doyle JJ (1992) Molecular Systematics of Plants. New York: Chapman and Hall.
53. Stern DB, Palmer JD (1984) Extensive and widespread homologies between mitochondrial and chloroplast DNA in plants. Proc Natl Acad Sci USA 81: 1946–1950.
54. Tanksley SD, Bernatzky R, Lapitan NL, Prince JP (1988) Conservation of gene repertoire but not gene order in pepper and tomato. Proc Natl Acad Sci USA 85: 6419–6423.
55. Tewari KK, Wildman SG (1966) Chloroplast DNA from tobacco leaves. Science 153: 1269–1271.
56. White EE (1986) A method for extraction of chloroplast DNA from conifers. Plant Mol Biol Rep 4: 98–101.
57. Wolfe KH, Li WH, Sharp PM (1987) Rates of nucleotide substitutions vary greatly among plant mitochondrial, chloroplast, and nuclear DNAs. Proc Natl Acad Sci USA 84: 9054–9058.
58. Young ND, Zamir D, Ganal MW, Tanksley SD (1988) Use of isogenic lines and simultaneous probing to identify DNA markers tightly linked to the Tm-2a gene in tomato. Genetics 120: 579–585.

4. Detection of DNA sequence variation for genome analysis

DONNA M. SHATTUCK-EIDENS, RUSSELL N. BELL &
TIMOTHY HELENTJARIS
Molecular Biology Group, Ceres/NPI, Salt Lake City, UT, USA

Introduction

Limitations of RFLP analysis

Genotypic analysis using RFLPs has proven to be a powerful tool for both plant geneticists and breeders, but it suffers from some limitations inherent the process. The entire procedure consists of many steps, some of which require sophisticated equipment and well-trained personnel. The analysis of low copy number sequences by Southern hybridization also operates too near the practical limit of detection (i.e., approximately 1 pg of actual target sequence in 5 μg of maize genomic DNA) and therefore can be unreliable. With RFLP studies involving large numbers of samples and many marker loci, the cost and time requirements become unacceptable. U.S. seed companies currently analyze an estimated five to ten million crop genotypes in the field every year; a significant fraction of these field-tested genotypes could instead be determined by RFLP analysis, if this alternative were practical. Although this is a tremendous opportunity to apply technology, existing RFLP processes are inadequate to meet the demands of such a volume of analyses. Using the best available technology in our laboratory, one technician can currently accomplish about 1000 genotypic analyses per year, where a single analysis consists of one genomic DNA preparation probed with 75 marker clones. Low throughput and the prolonged time to complete analyses are both significant obstacles to the wider application of this technology.

Tremendous differences in the level of variation detected by RFLPs have been noted among species [8]. In maize, approximately 95% of all 'unique sequence' clones, either cDNA or genomic, revealed RFLPs when tested with only three restriction enzymes against a selection of Corn Belt Dent germplasm. In addition, the average number of 'allelic' variants noted in this set of germplasm was greater than six per marker locus tested. In tomato, however, only 5–10% of the unique sequence clones were able to detect RFLPs within domesticated *Lycopersicon esculentum* germplasm and the average number of alleles revealed by those informative clones was only slightly greater than two. Like tomato, many important agronomic crop species, such as wheat and soybean, also display low levels of variability [21, 11] and this limited variability has severely hampered the extension of RFLP analysis to these species.

55

J.S. Beckmann and T.C. Osborn (eds) Plant Genomes: Methods for Genetic and Physical Mapping, 55–70.
© 1992 *Kluwer Academic Publishers.*

All of these problems have prompted a search for an alternative technology that would allow fast, low-cost analyses that could be applied to genotypic analysis of all species, including those with low levels of RFLPs.

Although RFLPs are a reflection of DNA sequence variation, they are a rather crude and indirect measure of this variation, since the polymorphism relies on altered restriction enzyme recognition sites or on significant size differences of inserted or deleted DNA between restriction sites. Due to this lack of resolution, many changes in DNA sequence such as single nucleotide changes or small insertions/deletions and rearrangements may go undetected in RFLP analysis.

We have considered alternatives to RFLP analysis capable of detecting minor DNA sequence polymorphisms (e.g., even single-nucleotide substitutions) and offering the advantage of greater sensitivity and higher throughput. The advantages of such an approach are illustrated by the results of an investigation of maize where more 'alleles' were revealed amongst a set of cultivars by direct DNA sequencing of small homologous regions (approximately 1000 bp) than were revealed by Southern analysis of much larger fragments (several thousand bp) [22]. Information gained from these investigations has been used to develop methods for detecting these sequence (allelic) differences by using the polymerase chain reaction (PCR) as will be described below.

We also examined melon, a species which exhibits relatively few RFLPs, in the hope that a similar strategy would reveal DNA seqence polymorphisms which are not detected by RFLP analysis. We considered the difference in occurrence of RFLPs between maize and melon to be a reflection of disparate incidence between the two species of significantly large insertions and deletions. We hypothesized further that despite this difference the frequency of single-nucleotide alterations would be comparable in maize and melon. Unfortunately, very little sequence variability was detected in melon: in a sequence of 1000 bp only two nucleotide differences were observed amongst individuals from a very diverse set of germplasm. In this case this comparison of sequence variation was no more informative than the RFLP analysis [22]. Obviously, no strategy of genetic analysis can be employed using marker loci if there is insufficient DNA sequence polymorphism in the surrounding regions. The loci used in the aforementioned investigation of maize and melon represent under-methylated and low-copy genomic regions. The successful identification of DNA sequence polymorphisms in recalcitrant species may require an examination of sequences representing other genomic areas such as the hypervariable regions found in humans and cattle [9, 5]. Some progress using this approach in plants has been reported [4, 19].

Alternative approaches to RFLPs

Nucleic acid sequences that differ by as little as a single base mismatch can be compared with techniques which employ the formation of duplexes containing one strand from each allelic variant. Mismatches resulting from imperfect homology can be detected either in RNA/DNA hybrids by cleavage with RNase A [17] or of DNA/DNA duplexes by chemical cleavage [3]. With the technique of denaturing gradient gel electrophoresis, differential melting has been exploited as a means of screening DNA duplexes for the presence of sequence heterogeneity [18]. In principle, these techniques, like RFLP, require no knowledge of the exact sequence differences that might exist among the alleles.

Sequence differences among alleles at specific loci can be determined by using separate clones representing the locus from each individual or by using DNA amplified from the individual plant genomes. To date most of the investigations of DNA sequence polymorphism have been achieved in the area of human genetics [see 13]. One method of testing for single-nucleotide mutations at specific loci makes use of allele-specific oligonucleotides (ASO) [12]. In this procedure the target DNA to be tested, after being immobilized onto a solid support, is hybridized with a labelled oligonucleotide that is complementary with the sequence overlapping the variable nucleotide position. Hybridization is performed under conditions in which the oligonucleotide will hybridize to all of the alleles. Multiple washes are then performed at successively higher stringencies causing mismatched combinations of probe and target to dissociate before perfectly matched combinations. In this way it is possible to determine whether the target sequence is identical to the oligonucleotide probe and thus make allele assignments.

Another approach for detecting DNA sequence polymorphisms is the oligonucleotide ligation assay (OLA) [14]. Here, the target sequences immediately surrounding the variable position are simultaneously hybridized with two adjacent oligonucleotides so that their junction occurs at the site of variation. If the sequence of the target is perfectly complementary with that of the oligonucleotides, they can be joined with DNA ligase. If the target DNA is mismatched at the junction, the ligase will fail to join the two oligonucleotides. By labelling one of the oligonucleotides, successful ligation can be visualized by high-resolution electrophoresis. An alternative to electrophoresis employs the coupling of biotin to the unlabelled oligonucleotide and assessing ligation by capture of the joined products on immobilized avidin. If the target is identical to the probes, the signal will be captured by the avidin, otherwise no signal will be captured. This plus/minus test allows discrimination of different alleles. Since nucleotide sequence determination is necessary for identifying these types of DNA sequence polymorphisms, investigators have found that, for some applications, direct sequence determination using efficient automated sequencers is the most readily applicable means of genotyping individuals.

The application of the procedures summarized above rely principally upon tremendous sensitivity for the detection and discrimination of target sequences. Requirements for such sensitivity are relaxed if the target sequences are amplified prior to detection. The most widely used method for the enzymatic amplification of specific sequences of DNA has been the polymerase chain reaction

(PCR) [7]. The PCR method permits the selective amplification, by a factor of 10^6 or more, of a specific segment of DNA in a complex mixture of sequences. The specificity is derived from information about the nucleotide sequence at the boundaries of the segment to be amplified. With this information, pairs of oligonucleotide primers are constructed, each with sequence complementarity with one of the ends. The target DNA is combined with a huge molar excess of these primers, all four deoxynucleoside triphosphates, and a heat-resistant DNA polymerase (*Taq*) from *Thermus aquaticus*. This mixture is heated to denature the target DNA then cooled to allow the primers to hybridize (anneal) with their complementary sequences. The DNA polymerase synthesizes new DNA from the primer/template complexes formed during the annealing step. The primers hybridize to the opposite strands of the target sequence and are oriented so that both newly synthesized strands extend across the interval to be amplified. Newly synthesized DNA, when denatured, also functions as template for subsequent DNA synthesis. Consequently, each iteration of the denaturation, annealing and extension steps, in theory, doubles the amount of DNA that lies between the primers, affording a logarithmic accumulation of DNA. Depending on the number of cycles of amplification, the yield of amplified DNA is sufficient to allow visualization without radioactive isotopes.

DNA amplification with PCR has been employed in human genetics for the identification of allelic variants that result from both size differences and nucleotide substitutions. PCR has been utilized for the sensitive genotyping of loci that possess minor differences in length with primers which flank the interval wherein polymorphism occurs [10]. In this approach the alleles correspond to various lengths of the PCR-amplified DNA. To increase the amount of information per assay, PCR has been extended to the simultaneous amplification of multiple intervals (multiplex PCR) [1]. Where nucleotide substitutions occur in certain restriction enzyme recognition sequences, the allelic variants can be distinguished by cleavage of hybridized oligonucleotide probes [20]. In this technique a pair of labelled oligonucleotide probes, each complementary with one of the alleles, are hybridized to the individual's amplified sequence. Matched duplexes can be cleaved by the appropriate restriction enzyme, whereas mismatched heteroduplexes will not be cleaved. By adjusting the sizes of the oligonucleotide probes, both homozygotes and heterozygotes can be distinguished by the size of the resulting cleavage fragments which are monitored by gel electrophoresis.

Procedures

Our strategy has been to use allele-specific primers in PCR amplifications to amplify and discriminate between polymorphic alleles. By adjusting conditions of the PCR and choosing oligonucleotide primer sequences appropriately, two allelic variants can be distinguished by PCR. The success of this technique relies upon the specific failure of PCR to amplify with primers that do not perfectly match the target sequences of heterologous allelic variants. If two homozygous individuals, A and B, differ in DNA sequence at a particular site, a primer can be made to match the heterologous region in individual A such that in a PCR reaction the A sequence will be amplified but the B sequence will not. Conversely, a primer can be made to match and amplify sequence B and not sequence A. In this way alleles A and B can be distinguished.

Basic PCR procedure

DNA extraction

Total genomic DNA was isolated essentially as described in Lassner *et al.* [15]. Approximately 0.5 g of fresh leaf tissue was crushed in a leaf squeezer (Ravenel Inc. Seneca, SC) with 0.8 ml of extraction buffer (100 mM Tris pH 7.5, 0.7 M NaCl, 10 mM EDTA, 10% CTAB, and 1% BME previously heated to 60 °C) and collected into a 1.5 ml microfuge tube containing 300 μl chloroform. The tube's contents were mixed well and incubated at 60 °C for at least 10 min but can go for up to one hour. The contents were transferred to a new tube and microfuged for 5 min. The aqueous supernatant was transferred to a new tube with 600 μl isopropanol, mixed well and microfuged for 5 min. The supernatant was decanted and the pellet washed in 800 μl of 80% ethanol. The tube was microfuged for 2 min, decanted, the interior blotted dry and the dried pellet was resuspended in 100 μl of 10 mM Tris pH 8, 1 mM EDTA. The yield of genomic DNA was approximately 10−40 μg depending upon the species and the condition of the tissue used. Successful variations of this procedure include the use of a SDS-based buffer (100 mM Tris pH 8, 50 mM EDTA, 1% SDS, 0.5M NaCl and 10 mM BME) as the denaturant instead of CTAB, substitution of Extractor column chromatography (Molecular Biosystems Inc., San Diego, CA) for the microcentrifugation steps and use of glass powder to bind the DNA and separate it from contaminating nucleases (Gene Clean, Bio 101, La Jolla, CA).

PCR reactions

PCR reactions were performed in reaction volumes of 25 to 300 μl composed of 10 mM Tris-HCl pH 8.4, 50 mM KCl, 2.5 mM MgCl$_2$, 10% DMSO, 0.2 mg/ml gelatin, 0.2 mM dATP, 0.2 mM dCTP, 0.2 mM dGTP, 0.2 mM dTTP, 0.1 μM primer 1, 0.1 μM primer 2, 0.2 μg/ml RNase A, 10 U/ml RNase T$_1$, 2 units/ml *Taq* DNA polymerase and 10 μg/ml genomic DNA. The input of DNA is such that there is about 1−10 amol or 10^5−10^6 copies of target sequence in the reaction. A mineral oil overlay of approximately 50−100 μl was used to prevent evaporation. Temperatures were cycled in a thermal cycler (TwinBlock System, Ericomp, San Diego, CA) with the following profile for ordinary amplifications: denaturation for 30 s at 93 °C, annealing for 60 s at 55 °C, and extension for 120 s at 72 °C. A portion (generally 30%) of the amplified product were sampled and electrophoresed on 1% agarose slab gels containing 0.5 µg/ml ethidium bromide. Thirty iterations of the profile described here were usually necessary to produce enough product for detection by agarose gel electrophoresis. Modifications of this procedure for specific primer pairs are described in the Results section.

Results

Initial experiments using this strategy showed some variability, but usually gave reduced yields in amplifications using mismatched primers where no product was expected. We set out to better define the parameters involved in discrimination, so that for any sequence, primers could be constructed so as to correctly discriminate mismatched sequences and amplify only the matching template.

Sequence-specific discriminating PCR could result from the diminished binding of imperfectly matched primers. Investigations of the hybridization of oligonucleotides to complementary DNA immobilized on a solid support indicate that internal mismatches have the greatest effect on hybrid stability [23]. Alternatively, a primer may not support amplification if, although bound to the target DNA, it is not extended by the *Taq* DNA polymerase. If the specificity of *Taq* DNA polymerase is such that a hydrogen-bonded nucleotide at the 3' end of the primer is required, a mismatch here may be more disruptive than mismatches elsewhere in the primer sequence. In order to design the most effective primers for sequence-specific amplification, we attempted to determine the relative consequence to amplification of internal and 3' mismatches of primers and substrates.

Figure 1 illustrates the effects of various mismatches in primer/template combinations. Genomic DNA from three inbred corn cultivars was used as template in a PCR reaction with three potentially discriminating primers, each 20 nucleotides in length (Fig. 1, lanes 1–9). One of the primers (BOL1) used in each of the reactions matches all three inbred templates, while the second primer in each reaction was chosen either to match the template DNA, to mismatch at an internal or 3' base, or to mismatch at both sites. In this way it was possible to assess the effect of amplification of an internal mismatch (at position 6 from the 5' end) or 3' terminal mismatch.

A considerable yield of amplified DNA is evident from the reaction in lane 1 of Fig. 1 where the primer MTB1, which exactly matches the MT template, is used in the amplification. MTB1 matches H99 at the internal site but the 3' terminal T is mispaired with another T. As can be seen from lane 2 there is less amplified product than was produced with the template MT. There is no detectable product in lane 3 where MTB1 mismatches the B73 template at both positions (C-A at position 6 and T-T at the 3' terminus). The presence of the

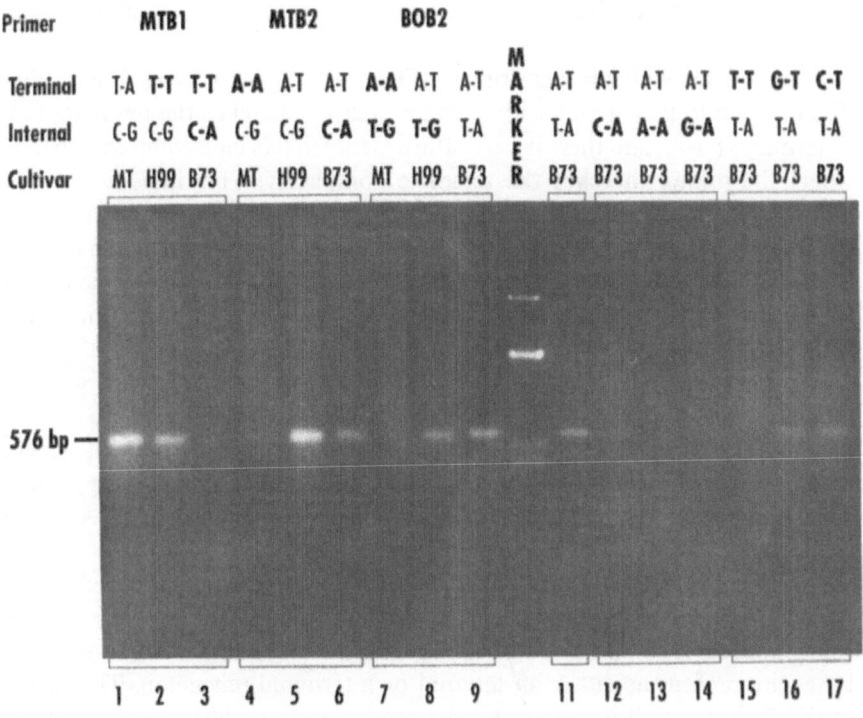

Fig. 1. PCR reactions were run for 30 cycles with graduated annealing temperatures starting with 62 °C. Three cycles were carried out at each degree down to 55 °C annealing, followed by nine cycles at 55 °C for a total of 30 cycles. Of each reaction 16 µl was electrophoresed through 1.5% agarose with 1 µg/ml ethidium bromide. Success of the PCR reaction was evaluated by the intensity of the stained band at the expected molecular weight. For each reaction, primer BOL1 was used in combination with the primer identified at the top of the figure: lanes 1–3 with MTB1, lanes 4–5 with MTB2 and lanes 7–9 with BOB2. Lane 10 contains the molecular weight markers. In lanes 11–17, primers were used to test the effects of nucleotide changes (C, A, G, or T) in two positions (internal or 3′ terminal). The nucleotides at the two positions are shown at the top of each lane (3′ terminal or internal). The first nucleotide of the pair is in the primer and the second is in the cultivar, mismatches are shown in bold type. The sequences of the cultivars and the first three primers are shown at the bottom of the figure. Dashes are used to indicate identical sequence.

second mismatch at the internal site accentuated the discrimination at this locus.

The products with another primer, MTB2, are shown in lanes 4–6 of Fig. 1. The reaction in lane 4 was with a primer that mismatches the template at the 3' terminus (A-A) and there is very little product in this lane compared to lane 5 where the primer matches the template completely. This indicates that the nature of the mismatch is important since in lane 2, a T-T mismatch at the 3' terminus did not reduce the product by as much as the A-A mismatch in lane 4. The template used in lane 6 is mismatched by this primer at an internal position (C-A) and the product is again reduced relative to lane 5, although not as much as in lane 4.

The primer-template combination with an internal G-T pair (Fig. 1, lane 8) amplifies almost as well as the perfectly matched primer (lane 9). In contrast, the mismatched primer (T-G internal and A-A terminal) fails to generate detectable product (lane 7). This is probably due to the terminal A-A mismatch, since the internal T-G had little effect alone (lane 8), and the terminal A-A alone reduced, but did not eliminate, amplification (lane 4).

A set of primers with various mismatches at one end and a 'fixed' primer, BOL1, were used to amplify a single inbred template, B73 (Fig. 1, lanes 11–17). The primer set is composed of variations of BOB2 such that all three alternative bases are present as either an internal or a terminal mismatch. The product of the perfect match is in lane 11, the products of amplification with internal mismatches C-A, A-A and G-A are in lanes 12, 13 and 14, respectively. In this case the product is reduced most when the mismatch juxtaposes two purines (either A-A or G-A). The products made from primers mismatched at the 3' end resulting in T-T, G-T and C-T mispairings are in lanes 15, 16 and 17, respectively. The G-T mismatch, lane 16, is effective in amplification as seen before. Interestingly, although both T-T and C-T are pyrimidine-pyrimidine pairs, T-T is more detrimental to amplification.

These experiments indicate that the nature of the primer's 3' base is important in accomplishing amplification. The most disruptive 3' mismatches occur where two purines or two pyrimidines are juxtaposed (e.g., A-G or T-T). Purine-pyrimidine pairs are much less disruptive and the G-T pair, although not normally found in DNA, is apparently extended by the polymerase perhaps because there is the potential for hydrogen bonding [24]. The data do not, however, exclude the possibility that sequence context or secondary structure of the template is involved in restricting primer performance.

In addition to the considerations outlined above, we have investigated other conditions to augment the discrimination between sequence variants. Optimal discrimination can be achieved when both amplifying primers are positioned at sites of sequence variation and are constructed so as to either both match or both mismatch the template (data not shown). We have also observed enhanced discrimination when elevated annealing temperatures are used in the PCR cycles. In the experiment in Fig. 1 a graduated annealing program was used in which the first three annealing steps were performed at 62 °C; the annealing temperature for subsequent sets of three cycles were sequentially dropped one degree down to 55 °C, whereupon the final nine cycles were annealed at 55 °C for a total of 30 cycles. In this way it was possible to enhance the discrimination, presumably because the matched primer anneals at a higher temperature than the mismatched primer and begins amplification several cycles before the mismatched primer. The advantage of such a graduated program is that it obviates the need for determining the appropriate annealing temperature for each pair of discriminating primers. Theoretically one program could suffice for any pair of primers, although in practice this may turn out not to be ideal.

The discriminating PCR amplifications described here were performed using a total of 30 cycles. This number of cycles is required for the production of enough DNA to be detected by electrophoresis with ethidium bromide staining. A hazard of discriminating PCR amplifications with this many cycles is the potential for amplifying spurious products that result from rare extensions of mismatched primers early in the process. For this reason it is important to perform the amplifications using an annealing temperature sufficiently low for matched primers to anneal yet too high for annealing of mismatched primers.

In order to estimate the optimal annealing temperature for PCR reactions, we determined the individual effective annealing temperatures for several primers in primer extension reactions. The experiment was set up to determine the effect of annealing temperature on the extension of a primer by *Taq* polymerase. The T_d of an oligonucleotide (the temperature at which one half of the hybrid dissociates) can be approximately calculated by summing 2 °C for every A-T pair and 4 °C for every G-C pair over the length of the primer [16]. The T_d calculated in this manner is useful for determining conditions for hybridization of an oligonucleotide to immobilized DNA in 0.9 M NaCl, however it is not directly applicable to hybridization in a PCR reaction. At the beginning of a PCR reaction to amplify a genomic sequence the target template is present at very low concentration. The primer extension experiment was set up to simulate the binding and extension of a primer in the later PCR reaction cycles, when the template is not limiting. Some of the same primers shown in Fig. 1 were examined in these experiments. The templates for these experiments were generated by amplifying the target DNA with two perfectly matched primers. The product was then used as template in the extension reaction. Several annealing temperatures were tested ranging from 55 °C to 72 °C in two-degree intervals. The template and primer combination was denatured for 5 min at 95 °C, lowered to the chosen annealing temperature for 30 s and extended by *Taq* DNA polymerase at 70 °C in the presence of [α-^{32}P]dCTP. The products of this single extension reaction were denatured and electrophoresed on polyacrylamide gels which were subsequently dried and autoradiographed with X-ray film. As the annealing temperature increased, the primers became less effective in the extension reaction presumably because they failed to anneal to the template. As predicted, the perfectly matched primer extended at a higher temperature than the mismatched primers.

Primer MTB 1

Temperature in °C

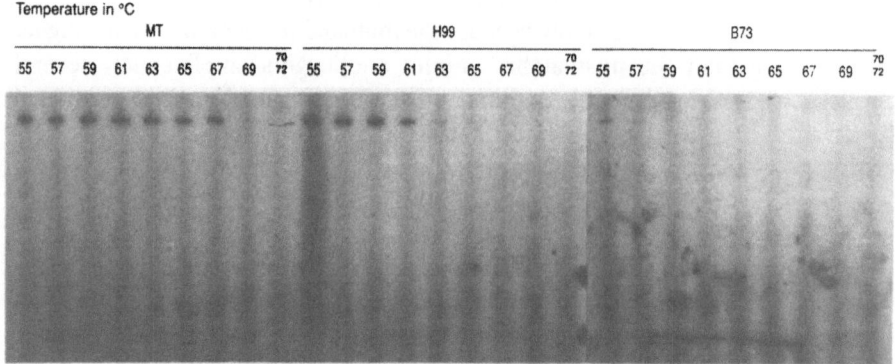

Fig. 2. The primer MTB1 (TCCTT CTATG TCACT GGCGT) was extended by *Taq* DNA polymerase on 50 ng of template amplified from cultivar MT, H99 or B73 as shown at the top of the figure. For each template, extensions were carried out after annealing at temperatures from 55 °C to 72 °C. The primer/template combination was denatured for 5 min at 95 °C, lowered to the chosen annealing temperature for 30 s, and the duplexes extended by *Taq* DNA polymerase at 70 °C in the presence of [α-^{32}P]dCTP. The products of this single extension reaction were then denatured and electrophoresed on polyacrylamide gels which were dried and autoradiographed with X-ray film.

The extension product from primer MTB1 is detectable at annealing temperatures up to 67 °C on the perfectly matched template, MT (Fig. 2). The same primer was tested on template H99, and the extension product was not detected above an annealing temperature of 61 °C (Fig. 2, compare to lane 2 of Fig. 1). MTB1 mismatches template H99 at the 3' terminus, aligning a T with a T. When MTB1 was tested on template B73 (two mismatches: C-A at position 6 and T-T at the 3' terminus) only a small amount of product was detected at the lowest temperature, 55 °C (Fig. 2). This demonstrates the effect of the single and double mismatches on the annealing of a primer 20 nucleotides long.

We went on to examine the effect of primer length by testing MTB1A, 18 nucleotides, MTB1B, 22 nucleotides and MTB1C, 14 nucleotides in length. The primers maintain the same 3' end, their difference in length being the number of nucleotides at the 5' end. The yield of extension on the matched template is comparable for 18, 20 or 22 nucleotide primers while the 14 nucleotide primer at annealing temperatures above 63 °C yields diminished extension product (data not shown). When these primers were tested on the mismatched template, length had a greater effect. The shorter primers are more affected by the mismatch than the longer primer. MTB1B (22 nucleotides) can be extended on template H99 at 63 °C, while on the same template MTB1A (18 nucleotides) is ineffective above 57 °C (see Fig. 3). MTB1C was not tested on this template. On the B73 template, which mismatches MTB1, MTB1A and MTB1B at two positions, none of the primers are extended at temperatures above 55 °C. MTB1C is only long enough to mismatch B73 at one site, the 3'

Primer MTB 1A

Temperature in °C

MT   H99   B⁻3

55 57 59 61 63 65 67 69 70   55 57 59 61 63 65 67 69 70   55 57 59 61 63 65 67 69 70

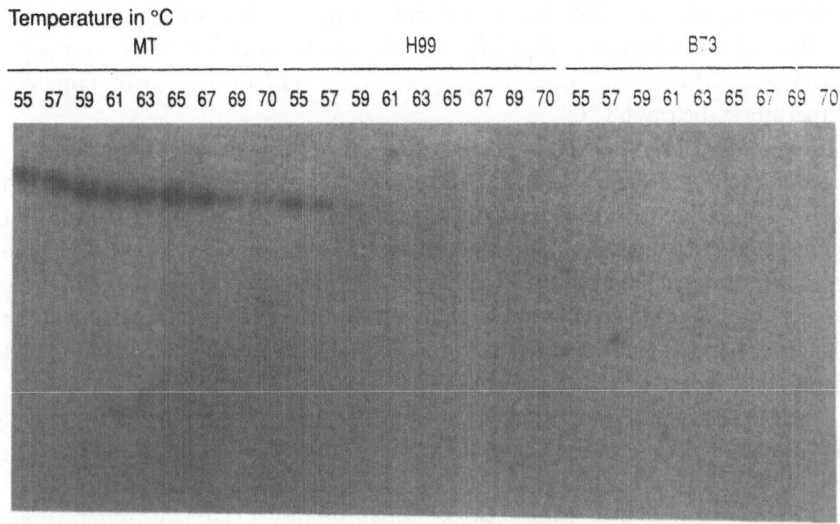

Primer MTB 1B

Temperature in °C

MT   H99   B⁻3

                         70   70   70

55 57 59 61 63 65 67 69 72 55 57 59 61 63 65 67 69 72 55 57 59 61 63 65 67 69 72

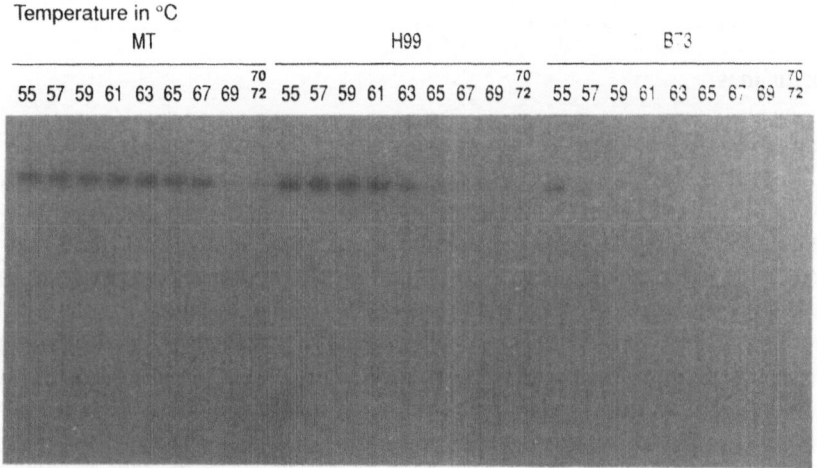

Fig. 3. Primers MTB1A (CTT CTATG TCACT GGCGT) and MTB1B (AATCCT CTATG TCACT GGCGT) were extended separately on templates amplified from cultivars MT, H99 or B73 as shown at the top of the figure. For each template extensions were carried out after annealing at temperatures from 55 °C to 72 °C. The primer/template combinations were denatured for 5 min at 95 °C, lowered to the chosen annealing temperature for 30 s and the duplexes extended by *Taq* DNA polymerase at 70 °C in the presence of [α-^{32}P]dCTP. The products of this single extension reaction were then denatured and electrophoresed on polyacrylamide gels which were dried and autoradiographed with X-ray film.

terminus, and it also fails to be extended above 55 °C (data not shown). The fixed primer which matches all of the templates, BOL1, was tested in a primer extension experiment. This primer extends at temperatures up to 68 °C (data not shown). This indicates that BOL1 is not the basis of discrimination when paired with MTB1 and used to amplify MT, H99 or B73 with annealing temperatures below 68 °C.

Improved discrimination is often realized in amplifications using elevated annealing temperatures, although the appropriate temperature depends on the pair of primers used and the nature of the mismatches in the discriminating primer. It is possible that 3' mismatches both diminish the stability of primer/template complexes and interfere with polymerase function. That mismatches at the 3' end of a primer disrupt amplification more than internal mismatches may reflect an attenuation of both annealing and extension, whereas internal mismatches lessen only hybrid stability.

These experiments suggest guidelines for designing primers and for choosing conditions for discrimination: (1) the primers should be short (about 14 nucleotides), (2) if possible, the region to be amplified should be chosen which allows for the placement of both primers at sites of variability, (3) the mismatch should be positioned at the 3' end of the primer, (4) if possible, two purines or two pyrimidines should be aligned in the mismatch configuration and (5) the annealing temperature should be high enough to inhibit extension from the mismatched primer but low enough to allow the matched primer to extend and the enzyme to remain active.

Conclusions

We have considered here an alternative technique for the genotypic analysis of plants that accommodates the requirements of high sample throughput and multiple (50–100) marker analyses. Much of current field analysis stresses the evaluation of segregating progeny for the possession of traits with either simple or complex inheritance, where a high fraction of the heritability may be due to relatively few genes. A reasonable near-term alternative technology for the laboratory is one capable of analysis of one to five marker loci with very high sample throughput. Such a system designed for a field station with its presently available facilities and personnel would significantly widen the application of genotypic analysis to plant breeding programs.

RFLP marker loci could be 'converted' to allele-specific primer sets by sequence comparison among a sample set of germplasm, using a process described herein and detailed in Shattuck-Eidens et al. [22]. This would allow for the identification of several alleles at any one locus, although detection would consist only of a 'plus-minus' type of evaluation. The markers most amenable to conversion to this type of analysis are those linked to genes of economic interest and for which large numbers of segregating progeny require screening. These marker sets could be used to follow the inheritance of simple traits such as recessive endosperm characteristics, disease resistance loci, or incompatibility factors; determine the presence of various loci involved in more complex traits such as yield and maturity; or test for seed purity amongst lots of seed.

We are presently developing a system that will allow a field station technician to genotype one to five loci in 100 plants per day. The isolation of genomic DNA from plant material and the processing for PCR analysis could all be done in a single day. The PCR reactions could be electrophoresed overnight in agarose gels containing ethidium bromide to visualize the results the next morning.

Developments elsewhere suggest that even this process can be further improved. Recently, allele-specific amplifications for the diagnosis of sickle cell anemia [25] and for genotyping human HPRT alleles [6] have been reported. Further improvements could incorporate a signal on the discriminating primer allowing the results to be read directly as described by Chehab and Kan [2]. In this chapter we have described only the amplification step of what is otherwise still a multistep process. Since the PCR reaction does not require the isolation of large quantities of high-molecular-weight genomic DNA, the usual DNA preparation procedures can be simplified. We are currently developing simplified and rapid DNA isolation techniques.

Acknowledgments

We gratefully acknowledge Jan Tivang for preparing the figures. This work was partially funded by NIH SBIR grant 1 R43GM40794–01.

References

1. Chamberlain JS, Gibbs RA, Ranier JE, Nguyen PN, Caskey CT (1988) Deletion screening of the Duchenne muscular dystrophy locus via multiplex DNA amplification. Nucleic Acids Res 16: 11141–11156.
2. Chehab FF, Kan YW (1989) Detection of specific DNA sequences by fluorescence amplification: A color complementation assay. Proc Natl Acad Sci USA 86: 9178–9182.
3. Cotton RGH, Rodrigues NR, Campbell D (1988) Reactivity of cytosine and thymine in single-base-pair mismatches with hydroxylamine and osmium tetroxide and its application to the study of mutations. Proc Natl Acad Sci USA 85: 4397–4401.
4. Dallas JF (1988) Detection of DNA 'fingerprints' of cultivated rice by hybridization with a human minisatellite DNA probe. Proc Natl Acad Sci USA 85: 6831–6835.
5. Georges M, Lequarré AS, Castelli M, Vassart G (1988) DNA fingerprinting in domestic animals using four different minisatellite probes. Cytogenet Cell Genet 47: 127–131.
6. Gibbs RA, Nguyen PN, Caskey CT (1989) Detection of single DNA base differences by competitive oligonucleotide priming. Nucleic Acids Res 17: 2437–2448.
7. Erlich HA, Gibbs R, Kazazian HH, eds (1989) Polymerase Chain Reaction. Current Communications in Molecular Biology. Cold Spring Harbor, NY.
8. Helentjaris T, King G, Slocum M, Siedenstrang C, Wegman S (1985) Restriction fragment polymorphisms as probes for plant diversity and their development as tools for applied plant breeding. Plant Mol Biol 5: 109–118.
9. Jeffreys AJ, Wilson V, Thein SL (1985) Hypervariable 'minisatellite' regions in human DNA. Nature 314: 67–73.
10. Jeffreys AJ, Wilson V, Neumann R, Keyte J (1988) Amplification of human minisatellites by the polymerase chain reaction: towards DNA fingerprinting of single cells. Nucleic Acids Res 16: 10953–10971.
11. Keim P, Diers BW, Palmer RG, Shoemaker RC (1989) Qualitative and quantitative studies of soybean with RFLP markers. In: Helentjaris T, Burr B (eds), Development and Application of Molecular Markers to Problems in Plant Genetics, pp. 35–40. Current Communications in Molecular Biology. Cold Spring Harbor, NY.
12. Kidd VJ, Wallace RB, Itakura K, Woo SLC (1983) α_1-Antitrypsin deficiency detection by direct analysis of the mutation in the gene. Nature 304: 230–234.
13. Landegren U, Kaiser R, Caskey CT, Hood L (1988) DNA diagnostics – techniques and automation. Science 229: 229–237.
14. Landegren U, Kaiser R, Sanders J, Hood L (1988) A ligase-mediated gene detection technique. Science 241: 1077–1080.
15. Lassner MW, Peterson P, Yoder JI (1989) Simultaneous amplification of multiple DNA fragments by Polymerase Chain Reaction in the analysis of transgenic plants and their progeny. Plant Mol Biol Rep 7: 116–128.
16. Meinkoth J, Wahl G (1984) Hybridization of nucleic acids immobilized on solid supports. Anal Biochem 138: 267–284.
17. Myers RM, Larin Z, Maniatis T (1985) Detection of single base substitutions by ribonuclease cleavage at mismatches in RNA:DNA duplexes. Science 230: 1242–1246.
18. Myers RM, Lumelsky N, Lerman LS, Maniatis T (1985) Detection of single base substitutions in total genomic DNA. Nature 313: 495–498.
19. Rogstad SH, Patton JC, Schaal BA (1988) M13 repeat probe detects DNA minisatellite-like sequences in gymnosperms and angiosperms. Proc Natl Acad Sci USA 85: 9176–9178.
20. Saiki RK, Scharf S, Faloona F, Mullis KB, Horn GT, Erlich HA, Arnheim N (1988) Enzymatic amplification of β-globin genomic sequences and restriction site analysis for diagnosis of sickle cell anemia. Science 230: 1350–1354.
21. Sharp PJ, Chao S, Desai S, Kilian A, Gale MD (1989) Use of RFLP markers in wheat and related species. In: Helentjaris T, Burr B (eds), Development and Application of Molecular Markers to Problems in Plant Genetics, pp. 29–34. Current Communications in Molecular Biology. Cold Spring Harbor, NY.

22. Shattuck-Eidens DM, Bell RN, Neuhausen SL, Helentjaris T (1990) DNA sequence variation within maize and melon: observations from polymerase chain reaction amplification and direct sequencing. Genetics 126 (In press).
23. Wallace RB, Shaffer J, Murphy RF, Bonner J, Hirose T, Itakura K (1979) Hybridization of synthetic oligodeoxyribonucleotides to ΦX174 DNA: the effect of single base pair mismatch. Nucleic Acids Res 6: 3543–3557.
24. Werntges H, Steger G, Riesser D, Fritz HJ (1986) Mismatches in DNA double strands: thermodynamic parameters and their correlation to repair efficiencies. Nucleic Acids Res 14: 3773–3790.
25. Wu DY, Ugazzoli L, Pal BK, Wallace RB (1989) Allele-specific enzymatic amplification of β-globin genomic DNA for diagnosis of sickle cell anemia. Proc Natl Acad Sci USA 86: 2757–2760.

5. Pulsed-field gel electrophoresis

RAYMOND VAN DAELEN & PIM ZABEL

Agricultural University, Department of Molecular Biology, Dreyenlaan 3, 6703 HA Wageningen, The Netherlands

General introduction

In view of the great impact of restriction fragment length polymorphism (RFLP) markers on both basic and applied plant research, considerable effort is currently being spent in constructing complete genetic linkage maps of the genomes of a wide variety of plant species. These maps should provide the framework for various types of applications, such as gene tagging and gene isolation. With markers for all chromosomes at intervals of 10–15 map units, the flow of any gene of interest through segregating generations in a breeding program can be monitored, using a linked RFLP marker as tag (for reviews see Beckmann and Soller [8], Helentjaris and Burr [57], Helentjaris *et al.* [58], Hille *et al.* [62], Landry and Michelmore [68] and Tanksley *et al.* [108]).

In some crop species, for which detailed linkage maps have been constructed, such as tomato and maize, this type of application ('RFLP-assisted breeding') is already becoming of age [108].

The second type of application involves the use of RFLP markers as molecular tools in cloning genes for which only a variant phenotype and map position is known. Nowadays, in principle, any such gene is clonable by an approach known as 'reverse genetics' when cloned molecular markers are available which flank the target gene on both sides at distances up to 2000 kb [48, 49, 86]. This very powerful approach emerged in the mid eighties, following the development of a variety of new techniques designed to allow the separation [94, 96] and cloning [18, 31, 81, 89] of DNA fragments in the size range encompassing the smallest distance (1 centiMorgan) covered by genetic recombinant analysis. Depending on the organism, this distance corresponds to 150–1000 kb. Since current electrophoresis techniques are capable of resolving DNA molecules of up to 10 megabases, the detection of one or more genetic loci along with flanking RFLP markers on a single restriction fragment becomes feasible. In human genetics, application of these long-range mapping and cloning strategies has already led to the construction of physical maps for large chromosomal regions (pseudo-autosomal region [17, 91], and HLA region of the major histocompatibility complex [55, 70, 71]), and to the molecular identification (Duchenne muscular dystrophy [20, 63, 112, 118]) and cloning of genes involved in hereditary disorders such as cystic fibrosis [30, 64, 81, 92, 93]. Such an undertaking would have been inconceivable five years ago.

J.S. Beckmann and T.C. Osborn (eds) Plant Genomes: Methods for Genetic and Physical Mapping, 71–100.

In plants, disease resistance loci are similar candidates for a reverse genetics approach as usually only their map position and mutant phenotype, but not their gene products, are known. Work is in progress in various groups to identify RFLP markers which are linked to genes conferring resistance to viruses, bacteria, fungi, nematodes or insects. For that purpose, an attractive strategy has been described by Young *et al.* [119] which circumvents the time-consuming and cumbersome RFLP analysis of segregating populations. By screening pools of genomic DNA clones on Southern blots of pairs of nearisogenic tomato lines differing for the presence/absence of the Tm-2a (tobacco mosaic virus resistance) gene and a small region of flanking DNA, clones tightly linked to the resistance gene could rapidly be distinguished by virtue of a restriction fragment length polymorphism between the isogenic lines. So far, this strategy has proven similarly successful in identifying RFLP markers very tightly linked to the *Fusarium oxysporum* resistance gene I_2 [94] and the root-knot nematode resistance gene Mi [65] in tomato. Given the availability of pairs of near-isogenic lines for a variety of genes, this approach should be widely applicable. An alternative strategy that would allow one to close in onto a gene of interest with bracketing markers has been recently discussed by Beckmann and Soller [9] but is beyond the scope of this chapter for a detailed discussion.

In moving from the flanking RFLP markers to the gene of interest, the molecular biologist has a number of tools at his/her disposal, including chromosome 'hopping' or 'jumping' [29, 30, 31, 88–90] and 'high altitude walking' [19, 82, see also Albertson *et al.*; Hauge and Goodman, this volume], which have been designed to boost the efficiency of standard chromosome walking [107]. Whichever of those strategies is going to be applied in cloning a particular gene of interest, an essential element in reverse genetics and physical mapping is the ability to prepare and fractionate DNA molecules at the megabase level. In this chapter we will focus on methods originally developed by Schwartz and Cantor [95] for this purpose and which are also applicable to plant systems [36, 44, 45, 52, 113, 115]. Several other reviews on Pulsed-Field Gel Electrophoresis have also been published [2, 3, 22, 101, 104].

Pulsed-field gel electrophoresis

Development of pulsed-field gel electrophoresis systems

In conventional (single-field) electrophoresis techniques, separation of DNA molecules is limited to those smaller than 25–50 kb. Above this size, all molecules migrate with the same mobility. Although separation of much larger molecules (up to 700 kb [40]) has been reported, the poor resolution and the use of very low-percentage, fragile agarose gels render this technique inadequate for most purposes.

The gel electrophoresis technique developed by Schwartz *et al.* [95, 97], termed Pulsed-Field Gradient Gel (PFGG) electrophoresis, is fundamentally different from the conventional electrophoresis techniques in that two, perpendicularly oriented electric fields are applied in alternating fashion. Thus, separation of DNA molecules up to 2000 kb in size could be achieved. Many modifications, all based on the same concept of using two electric fields applied alternatingly, were soon introduced to extend the separation range up to 10,000 kb. Here, we will discuss the various Pulsed-Field Gel (PFG) electrophoresis systems that have been developed over the past 5 years, with special emphasis on Field Inversion Gel Electrophoresis (FIGE) which is employed in our laboratory.

In the original PFGG electrophoresis method [95] both a homogeneous and an inhomogeneous electric field are applied which are generated by an array of electrodes placed around the gel as indicated in Fig. 1. Separation is accomplished by repeatedly switching the perpendicularly oriented electric fields at specified time intervals (the pulse times), over a certain period. The idea behind this approach was based on the viscoelasticity studies of Klotz and Zimm [66], implying that molecules which have been aligned in the direction of an electric field, require some time to reorient themselves before becoming mobile again when subjected to a change in the direction of the electric field. The time needed for reorientation of a DNA molecule was called relaxation time and is supposed to depend on the size of the molecule. Since large molecules need more time to reorient themselves than smaller molecules, they have a lower net mobility. Thus, the rate of migration is based on the rate at which DNA molecules alter their shape within the gel matrix. All molecules too large to be reoriented within the applied pulse times show the same mobility and combine to what is called the compression zone (also referred to as limiting mobility). The mechanism described above and the models subsequently proposed [23, 67, 106] do not explain all the features observed in PFG electrophoresis. Recently, a model has been proposed [34, 35] which seems to provide a better explanation and is moreover supported by studies in which the movement of DNA molecules has been visualized while undergoing electrophoresis [96, 103]. This model is termed geometration.

Due to the particular configuration of the electric fields in the Schwartz and Cantor setup, separation of DNA molecules occurs in diagonal tracks across the gel, making size estimates rather difficult. Since this is a serious impediment, alternative systems have been developed.

The same year a modification was published by Carle and Olson [24, 25] which was named Orthogonal Field Alternation Gel Electrophoresis (OFAGE). By using two orthogonal, inhomogeneous fields (Fig. 1), with the gel at an angle of 45°, a symmetrical separation pattern was obtained with straight lanes in the center of the gel. These improvements made size estimates more accurate, though there were still some flaws: lanes at the sides are still strongly curved and 'sidestepping' (samples from one slot run off from the main track and so form a 'shadow' [106]) also occurs. An advantage of the OFAGE system, however, is that, due to the inhomogeneous fields [100], bands usually

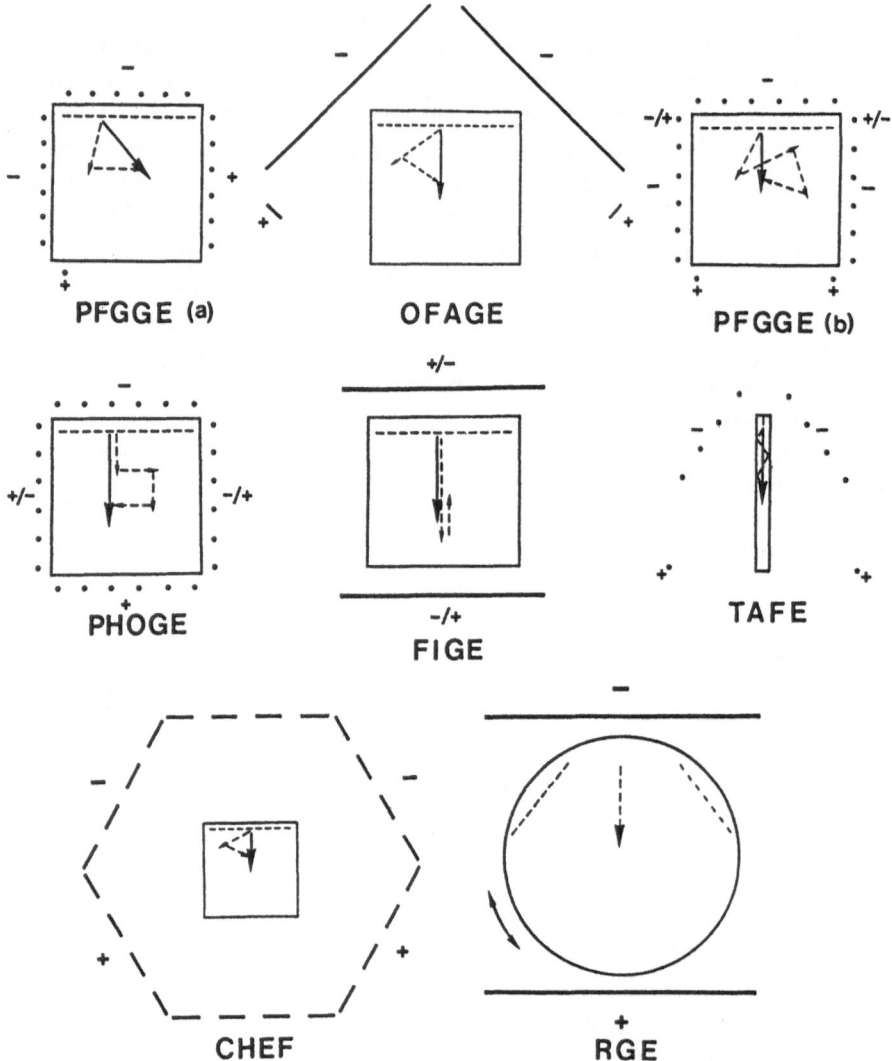

Fig. 1. Schematic representation of the various PFG electrophoresis systems. The PFGGE system is shown in its original ((a); Schwartz and Cantor [95]) and modified ((b); McPeek *et al.* [79]) version. Actual migration paths of the DNA molecules are indicated as dashed arrows and the resulting net migration with solid arrows. Drawings are not to scale. For abbreviations see text.

are sharper than in systems in which homogeneous electric fields are employed.

A modification of the PFGG electrophoresis system, with nearly the same electrode configuration but with a more sophisticated switching pattern consisting of inhomogeneous electric fields varied in 4 different directions, has been developed by McPeek *et al.* [79] (Fig. 1). Results were, however, only slightly better than with the OFAGE system.

With the introduction of Field Inversion Gel Electrophoresis (FIGE [23], Fig. 1) the problem of curved lanes was overcome. Another advantage of FIGE is its simple construction. Using a standard horizontal electrophoresis apparatus, in conjunction with a switching unit to control the electric field, we have obtained good resolution in the 20 to 1500 kb range [113]. Results by Turmel and Lalande [109] show that FIGE is also capable of separating the *Schizosaccharomyces pombe* chromosomes (3.5, 4.6 and 5.7 Mb in size [39]).

In FIGE, one homogeneous electric field is applied and its polarity is reversed periodically. Net migration in the forward direction is obtained by applying a forward pulse that is longer in duration than reverse, usually at a ratio of 3:1. Alternatively, net forward mobility can be achieved by applying different field strengths for the forward and reverse pulses [50]. A combination of different field strengths and pulse times for forward and reverse electrophoresis has been applied by Lalande *et al.* [67] to achieve separation over a wider range in relatively short runs.

A peculiar phenomenon observed with FIGE is that when pulse times of constant length were used, molecules above a certain size (depending on the length of pulse times used) show an increased mobility (see also sections on electrophoresis parameters). This can be overcome, to a large extent, by lengthening the pulse time over the course of the electrophoresis run, a process called 'time ramping'. Another, though usually not so serious, problem with FIGE is the relatively poor resolution over a larger size range. Also bands are usually more diffuse than in other systems.

Two-dimensional FIGE systems have been described in which entire lanes from the first FIGE run were subjected to a second round of FIGE separation – using different parameters – perpendicular to the first [7]. With this approach, the speed of constructing restriction maps in organisms with small genomes could be increased as only ethidium bromide-stained gels have to be analyzed without subsequent hybridization analysis [7].

Another PFG variant in which inhomogeneous fields are used, is Transverse Alternating Field Electrophoresis (TAFE [46]). In this system, the gel is placed vertically with the electrodes on both sides of the gel, as indicated in Fig. 1. Thus, homogeneous fields are obtained within the gel that have the same strength over the entire width of the gel, even though the electric field strength varies from top to bottom. The accomplished separation is similar to the systems developed before. Disadvantages are that small molecules (less than ca. 25 kb) are lost and only small gels can be run. Scaling up, to allow the use of 20 × 20 cm gels, requires much larger volumes of buffer [3].

Chu *et al.* [27] applied the principles of electrostatics to gel electrophoresis for construction of their PFG apparatus. By their method, Contour Clamped Homogeneous Electric Field (CHEF) gel electrophoresis (Fig. 1), two (homogeneous) electric fields under an angle of 120° are created, using a hexagonal contour of electrodes, clamped to set electric potentials. With this apparatus a very good resolution (separation of all yeast chromosomes in one run) and straight lanes across the entire width of the gel are obtained.

With the Rotating Gel Electrophoresis (RGE, Fig. 1), or Crossed Field Gel electrophoresis of Southern *et al.* [106], results similar to the CHEF system were obtained. In RGE, one homogeneous electric field is applied and the reorientation of the DNA molecules is obtained by periodically rotating the gel. Other rotating systems have been described by Gemmill *et al.* [47] and Anand [3]. These rotating gel systems have the advantage that the reorientation angle can be varied by simply changing the angle of rotation, thus determining the speed and range at which separation is achieved. Drawbacks are that very short pulse times cannot be used and, consequently, separation of molecules smaller than 50 kb is not possible; in addition, these mechanical devices are more sensitive to wear.

In the Pulsed Homogeneous Orthogonal field Gel Electrophoresis (PHOGE [5]) system (Fig. 1), migration of DNA molecules proceeds in homogeneous electric fields placed perpendicularly to one another. The usually poor resolution associated with the use of electric fields at an angle of 90° [27], is more or less overcome by applying a particular switching regime, in which molecules undergo 4 reorientations per cycle instead of only two. Though the PHOGE system provides a better separation and resolution than OFAGE, it appears worse than CHEF, and in addition, DNA tracks are not completely straight.

The youngest offspring in PFG systems is PACE [28] or Programmable, Autonomously Controlled electrode gel Electrophoresis. With this system all electric field parameters (number and angle of electric fields, voltage gradients, pulse time and time ramping) are adjustable and many types of pulsed-field gel electrophoresis can be chosen (OFAGE, FIGE, CHEF, PHOGE).

Currently, it is difficult to say which system gives the best results as many are still being perfected and new systems are being developed (e.g. Electrophoresis Device (ED) by David Schwartz [98]). Our experience lies mainly with the FIGE system, which is a very cheap and convenient system, especially for separation of DNA less than 1500 kb.

Parameters in pulsed-field gel electrophoresis

Although all electrophoresis parameters are discussed separately, one has to keep in mind that changing one parameter can have a strong influence on the effect of other parameters. For instance, separation obtained at 6 V/cm and pulse times of 90 s is almost similar to that at 3 V/cm with 300 s pulse times [13], while changing the temperature from 4 to 13 °C has an effect comparable with lengthening the pulse time from 70 to 90 s [13]. Therefore, applied conditions are always a compromise between the various parameters. Since FIGE has some very peculiar characteristics, it will be dealt with separately.

Parameters in PFG electrophoresis
Pulse time. In PFG systems, the most important parameter to affect the separation range is the pulse regime [13, 26, 76, 102, 106, 114]. In general, larger molecules are better separated when longer pulse times are applied. As a rule

of thumb, it holds that to obtain as large a resolution as possible over the desired separation range, pulse times as short as possible should be used (e.g. when very accurate size determinations are important), whereas long pulse times are required to obtain separation over as large a range as possible (at loss of resolution). At long pulse times, however, separation of smaller molecules decreases and molecules become compressed into one band, thus forming a compression zone below the separation zone. In Table 1 some pulse regimes with corresponding separation ranges for a CHEF system are presented.

Table 1. Pulse times with separation ranges and voltages when using a CHEF DRII (BioRad) gel box

Pulse time	Separation range	Voltage
1–10 s	< 100 kb	200 V
50–100 s	100–200 kb	150–200 V
120–180 s	2000–4000 kb	125–175 V
10–60 min	> 4000 kb	50–75 V

To obtain high resolution over wider separation ranges in a single run, different pulse times, or time ramping, should be applied.

Temperature. In general, the temperature during electrophoresis is maintained at 10 to 15 °C. At higher temperatures, bands become blurred [12, 75]. On the other hand, DNA is more mobile at higher temperatures [116] (25 to 35 °C) and the fractionation range of large molecules increases. Snell and Wilkins [104] have applied this effect to separate *Candida albicans* chromosomes, estimated to be at least 5 to 10 Mb in size. By using temperatures higher than 25 °C (but less than 35 °C, as resolution then becomes too weak) in combination with pulse times of only 80 to 120 s, those chromosomes could be clearly resolved in runs of approximately 1 day.

Agarose concentration. For most purposes 1% (w/v) gels are sufficient. With increasing agarose concentration, the mobility of DNA molecules decreases and, as larger molecules appear to be more strongly affected, the separation range is decreased. Resolution, however, is improved [12, 75]. At agarose concentrations of more than 2% (w/v) separation is very poor. For fractionating very large molecules (e.g. *S. pombe* chromosomes) sometimes gels of only 0.5% agarose [114] are used, although this does not seem to be crucial since *C. albicans* chromosomes can be separated in gels of 1.5% agarose [104]. As already mentioned, the effect of increased agarose concentration can be completely or partially reversed by using longer pulse times or higher temperatures.

Type of agarose. No systematic studies have been performed dealing with the effect of the agarose type. Different kinds of agarose, however, can influence the electrophoretic behaviour of DNA molecules [110]. To increase the speed of separation of very large molecules, special types of agarose have been developed which possess a higher gel strength and allow low-percentage gels to be made with high mechanical strength.

Electric field strength. For separating DNA molecules in the range of 1 to 2000 kb, 5 to 7 V/cm electric fields are usually applied. At higher voltages the separation range is larger, as is the absolute mobility [12, 76]. The resolution, however, is much better at not too high (less than 10 V/cm) field strengths. Very large molecules should be separated at low field strengths (1 to 3 V/cm), as shown by the separation of *S. pombe* and *Neurospora crassa* chromosomes at field strengths of 1.3 and 2 V/cm, respectively. These low voltages, on the other hand, imply that long runs are required (as long as 7 days for *N. crassa* chromosomes). Field strengths higher than 10 V/cm result in poor recovery and smeared bands [76].

Reorientation angle. At reorientation angles of 90°, the resolution of DNA molecules is poor [27, 95]. In the original designs of Schwartz and Cantor [95] and Carle and Olson (OFAGE [24]), respectively, reorientation angles were larger than 90° (due to the inhomogeneous fields), even though the setting of the electrodes was at an angle of 90°. Studies in which angles of 60°, 90° and 120° were applied [27], revealed that separation was superior at a 120° angle. No major differences in separation occurred between 105° and 165° [12, 21, 106]. However, at a reorientation angle of 180°, as applied in FIGE, some interesting features were observed (see FIGE section). The speed at which separation is attained, differs greatly with the applied reorientation angle. When this angle is reduced, the velocity at which separation is obtained increases. With respect to the separation of *S. pombe* chromosomes, Clark *et al.* [28] reported a 3.2-fold decrease of separation time when 96° was applied as compared to an angle of 120°.

DNA topology. The electrophoretic behavior of circular and linear DNA molecules is completely different in PFG systems. Several studies concerning the mobility of circular molecules in PFG systems have been published [10, 61, 74, 99, 105]. From these it becomes clear that many parameters (electrophoresis system, pulse time, agarose concentration, temperature and even the number of superhelical turns in supercoiled molecules) have an influence on mobility. Generalizations are therefore as yet difficult to make. Some patterns, however, begin to emerge: circular molecules are much less mobile in PFG systems than linear molecules. Supercoiled circular molecules, in the range from 4 to 16 kb, are resolved over the same area as yeast chromosomes with sizes ranging from 240 to 1500 kb (in OFAGE [61]). The supercoiled molecules are more mobile than the relaxed circular molecules which in their turn are somewhat more mobile than nicked open circular molecules [61]. Unlike linear molecules, their relative mobility is not, or only slightly (depending on specific electrophoresis conditions and size) affected by pulse time (at least not for molecules up to 85 kb) [10, 61, 74, 99, 105]. There is, however, an effect on absolute mobility. Supercoiled molecules smaller than 50 kb show the lowest mobility at 10 s pulse times (as compared to both shorter and longer pulse times [99, 105]). The pattern for nicked and relaxed circular molecules is not so clear. Beverly [10] reports that very large (30 and 85 kb) relaxed circular molecules do not enter OFAGE gels, whereas supercoils do.

Parameters in field inversion gel electrophoresis
A striking feature of FIGE is that when pulse times are kept constant the mobility of DNA goes through a minimum as a function of size. For example, at forward and reverse pulse times of 15 and 7.5 s, respectively, molecules of approximately 650 kb show a very low mobility and enter into what is called a minimum mobility zone, whereas both smaller and larger molecules are more mobile [23]. As this phenomenon seriously disturbs accurate size determinations, conditions have been sought for to avoid this. According to Carle *et al.* [23] this effect should be avoided by applying a time ramp (a gradual increase of pulse time) throughout the duration of the electrophoresis run. In common with Lalande [67], Ellis [36] and Bostock [16], we found, however, that the use of time ramps does not preclude the formation of this minimum mobility zone. The compression zone, similar to other PFG sytems, is located well below the minimum mobility zone. All in all, six different zones can be distinguished [16]: a zone in which small molecules (less than 20 kb) are separated as in conventional electrophoresis (I), a lower compression zone (II), a separation window, in which mobility decreases linearly with molecular weight (III), a minimum mobility zone (IV), a zone of reversed mobility (increase of mobility with molecular weight, V), and an upper compression zone (VI). The most important zone is the separation window in which mobility decreases linearly with increase in size. So, as to avoid any ambiguity over size estimates, a pulse regime should be applied that allows separation of molecules larger than the ones studied.

Pulse times. Like in other PFG systems, the pulse time (time ramps) is the major factor influencing the separation range [14, 16, 33, 50, 59, 60]. Using time ramps which encompass short pulse times, small molecules are separated (at great resolution). For separation within the range from 1 to 50 kb, asymmetric voltage FIGE [14, 15] performs best as compared with other PFG systems. When time ramps with long pulse times are used, the separation range is extended to the higher-molecular-weight molecules (at loss of resolution, however). To obtain separation over a large range of molecular weights and to preserve a good resolution, a compromise is to apply several time ramps in one run [16, 113]. In Fig. 2 (A and B) an example is given of a run comprising three different time ramps. For separation over 1500 to 2000 kb, FIGE is not the best choice. In Table 2, some time ramps with corresponding separation ranges are given.

Table 2. Electrophoresis conditions and corresponding separation ranges in FIGE, using either constant (C) or ramped (R) pulse regimes (mode) at a forward/reverse ratio of 3. When voltages and pulse times for forward and reverse pulses differ, both are given.

Pulse times	Mode	Duration (hours)	Field strength (V/cm)	Range (Kb)	Source
50/20 ms	C	–	6	1–25	[14]
50/20 ms	C	–	9/6	1–50	[14]
1.5–4 s	R	–	7.5	10–100	[16]
1–6 s	R	16	ca. 7	ca. 10–200	*
1–6 s	R	16	ca. 5.8	10–150	*
1–20 s	R	16	ca. 7	10–350	*
1–20 s	R	16	ca. 5.8	10–250	*
1–40 s	R	16	ca. 7	10–1000	*
1–60 s	R	16	ca. 7	50–1600	*
1–60 s	R	16	ca. 5.8	50–800	*
6–75 s	R	22	ca. 7	50–2000	*
3000/1200 s	C	80	1.3/0.5	2000–6000	[109]

* Results from our own research.

Forward/reverse ratio. The ratio between forward and reverse pulse times is usually kept at 3, which turns out to give the simplest separation pattern [59]. At a higher ratio, the separation range increases, but with a loss of resolution [16]. More complicated patterns (several zones of limiting mobility) emerge upon using either lower or higher ratios.

Temperature. In FIGE, the effect of temperature on separation is similar to that in PFG electrophoresis. An increase in temperature leads to an increase in absolute mobility, while the separation zone is extended to the higher molecular weights [16]. Olschwang and Thomas [84] have made use of this temperature effect by applying a temperature gradient over the gel to improve resolution.

Agarose concentration. Effects of different agarose concentration are also similar to the effects observed in PFG electrophoresis. At higher concentration, bands are sharper and separation is limited to lower molecular weights.

Electric field strength. Variations in the electric field strength in FIGE have effects similar to PFG. Increasing the field strength leads to an increased separation window, however, at loss of resolution as bands tend to become more diffuse. The use of very high field strength (> 10 V/cm) gives very diffuse bands [16]. For separation of large molecules, low voltages are required [109]. When separation of very small molecules (1 to 50 kb) is required, forward field strength should be larger than reverse field strength [14, 15] (see also Table 2).

DNA topology. Only a few studies [72, 105] concerning the behavior of circular molecules in FIGE have been published. Supercoiled molecules (2 to 16 kb in

size) are less mobile than linear molecules of similar size (at least in gels of 0.8%
agarose, or more [105]). At an agarose concentration of 1.0% the mobility is
pulse-time-independent, whereas at both higher and lower concentrations the
mobility is pulse-time-dependent [105].

Open circular molecules (in the range from 2.9 to 56 kb) are mobile in FIGE
gels [72], in contrast to OFAGE [11], and show a mobility (at short pulse times
of ca. 1 s) only slightly less than linear molecules [72]. We have found that
intact (supercoiled?) tomato chloroplast molecules (152 kb) are not mobile in
FIGE at field strengths of approximately 7 V/cm and pulse times in the range
from 1 to 150 s [113] (see also Fig. 2D).

Procedures

Preparation of high-molecular-weight DNA

Introduction

The essence of preparing DNA of very high molecular weight that is several mega-bases in size is to avoid shearing by mechanical stress and nuclease activity during cell breakage and subsequent DNA purification. This is achieved by suspending cells in molten agarose and, after solidification, carrying out all steps of DNA purification on the agarose samples. In the original protocol by Schwartz and Cantor [95] agarose blocks, referred to as inserts and later also as plugs, were made. An alternative procedure involves the preparation of agarose beads [32] in which the cells are embedded. As this procedure requires vigorous vortexing and plant protoplasts are rather fragile, we have not employed this procedure. The embedded cells are lysed by incubating the agarose plugs in a solution containing a detergent (either sodium dodecyl sulfate or N-lauroyl sarcosine, sodium salt), proteinase-K and a high concentration of a chelating agent (0.5 M EDTA). The large DNA molecules released from the cells remain trapped within the agarose matrix, whereas the rest of the cell material is degraded and diffuses out.

The following procedure has been developed for tomato protoplasts, but should be applicable to other plant species. A procedure for the isolation of tomato protoplasts has been described elsewhere [113]. It should be emphasized that only protoplast batches containing at least 95% viable cells (as determined by fluorescein diacetate staining [69, 117]) are to be used as otherwise a large fraction of DNA is degraded [113]. For the preparation of yeast DNA and phage lambda DNA-concatemers, which are used as markers, special procedures will be described. Procedures for preparing high-molecular-weight DNA from bacteria and other organisms have been described by Smith et al. [102].

Solutions used for preparing high-molecular-weight DNA should all be autoclaved and instruments that come in contact with agarose plugs should be sterilized.

Steps in the procedure

1. Prepare protoplasts according to [113] and resuspend the protoplasts in wash medium (CPW salts [42] consisting of 27.2 mg/l KH_2PO_4, 0.16 mg/l KI, 0.025 mg/l $CuSO_4 \cdot 5H_2O$, 0.101 g/l KNO_3, 0.246 g/l $MgSO_4 \cdot 7H_2O$ and 2% (w/v) KCl) at a concentration of 50 million per ml, so as to obtain enough DNA per plug to allow the detection of single-copy sequences in Southern blot hybridization experiments. For tomato this amounts to 4 to 5 μg per plug, assuming a haploid genomic weight of 0.74 pg [43] and a plug volume of 130 μl.

2. Warm the protoplast suspension (in wash medium), to 37 °C and add to the suspension an equal volume of 1% (w/v) low-melting-point (LMP) agarose (Seaplaque, FMC) made in protoplast wash medium, at 37 °C and mix gently.

3. Pipet the agarose-protoplast mixture into a prechilled plug-former, using a 1 ml plastic pipet tip, of which the tip has been cut off, and allow the agarose to

solidify for 15 to 30 min on ice. A mold for the agarose plugs can be easily made by cementing together perspex strips with slot-sized notches [111]. The molds should be thoroughly cleaned with alcohol before use and covered with tape on one side.

4. Push the solidified plugs out of the mold into the lysis mix, consisting of 0.5 M EDTA (pH 8.0), 1% (w/v) N-lauroyl-sarcosine and 1 mg/ml proteinase-K. When handling the plugs, no metal instruments should be used as metal ions may induce strand breaking [102].

5. Incubate the plugs in lysis mix (approx. 2 ml per plug) for 48 h at 50 °C. Change the lysis mix once, after 24 h. When proper lysis occurs, the plugs become clear within a few hours while the lysis mix turns dark green. To render as much DNA as possible accessible to restriction digestion, a prolonged (48 h) incubation in lysis mix is necessary [113].

6. After lysis, plugs can be stored in fresh lysis mix at 4 °C for prolonged periods (at least up to 1 year) without noticeable degradation.

Restriction digestion

Introduction

As in PFG systems the size range over which separation can be obtained is much larger than in conventional electrophoresis systems. Restriction enzymes should be selected that generate DNA fragments in the appropriate size ranges. Some enzymes are particularly suited as they recognize 8 bp sequences which also contain one or several CpG dinucleotides. In mammals [12] and plants [4] this dinucleotide is under-represented and often methylated [4, 12, 51]. Two such enzymes are currently known: *Not*I (GCGGCCGC) and *Sfi*I (GGCC(N)5GGCC). An 8 bp sequence consisting only of A's and T's (TTAATTAA) is recognized by *Pac*I. Other enzymes, that recognize either 7 bp sequences like *Rsr*II (CGG(A/T)CCG), or 6 bp sequences containing one or more CpG's in their recognition site, are also effective in generating large restriction fragments. In Table 3 a list of appropriate restriction enzymes is presented.

By specifically methylating the target DNA and subsequent digestion with a selected restriction enzyme additional cleavage specificities may be generated [83]. McClelland *et al.* [77] applied this approach to generate site-specific cleavage at 8 and 10 bp sequences, using a combination of M.*Taq*I and M.*Cla*I respectively, prior to restriction digestion with *Dpn*I. This technique has been applied by Bernards *et al.* [10] to agarose-embedded DNA.

Upon hybridization, sometimes multiple bands can be observed, even after incubation with an excess of enzyme for prolonged periods. These 'natural partial' digestions are due to partial methylation of the cytosine in the CpG base pairs [51, 78, 91]. These partials can be very helpful, especially in constructing long-range restriction maps [6, 18, 30, 38, 71, 80].

The following protocol is for a complete restriction digestion. Partial digestions, which often have to be performed for the construction of long-range restriction maps, are usually carried out by incubating agarose plugs with small amounts of enzyme (0.1 to 10 units) and for different times (varying from 15 min to 1 h). For each restriction enzyme we usually perform a test series to determine the optimum conditions. An alternative is to control the digestion by using either limiting Mg^{2+} concentrations [1], or using excess restriction enzyme in the presence of excess methyltransferase [53, 54].

86

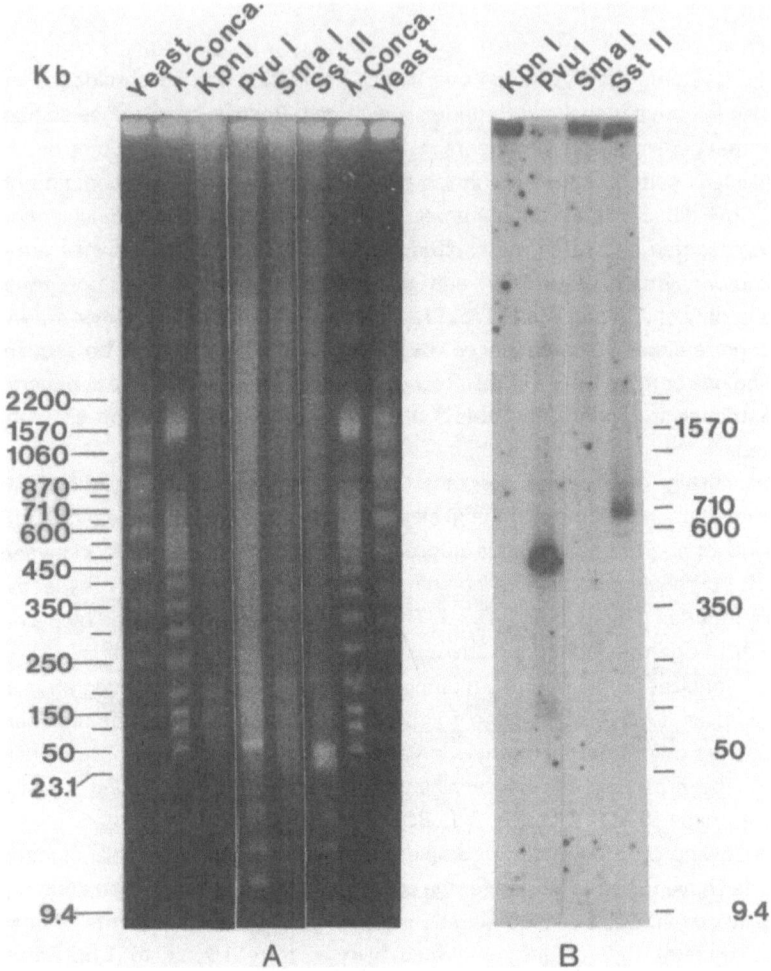

Fig. 2. Restriction digestion and hybridization analysis of lysed agarose-embedded tomato proto-plasts.

A and C. Agarose-embedded protoplasts were lysed for 48 h and the released DNA was digested with the restriction enzymes indicated and electrophoresed using a sequence of time ramps of 1–20 s (8 h), 3–65 s (14 h) and 5–150 s (20 h), respectively, at 5.8 V/cm. The ethidium bromide-stained FIGE gels are shown. Note the absence of specific chloroplast restriction bands (in the size range from 9 to 100 kb) in the *Kpn*I and *Sma*I digests of gel A, indicating a poor digestion.

B. Autoradiograph (1 day exposure, one intensifying screen) of the blotted gel shown in A, after hybridization with a single-copy probe (*Adh2* of tomato). Note the absence of specific restriction fragments in the *Kpn*I and *Sma*I digests.

D. Autoradiograph (15 min exposure, one intensifying screen) of the blotted gel shown in C, after hybridization with total nick-translated tomato chloroplast DNA as probe. Note that all intensely ethidium bromide-stained restriction bands show up in the autoradiograph, indicating their chloroplast origin.

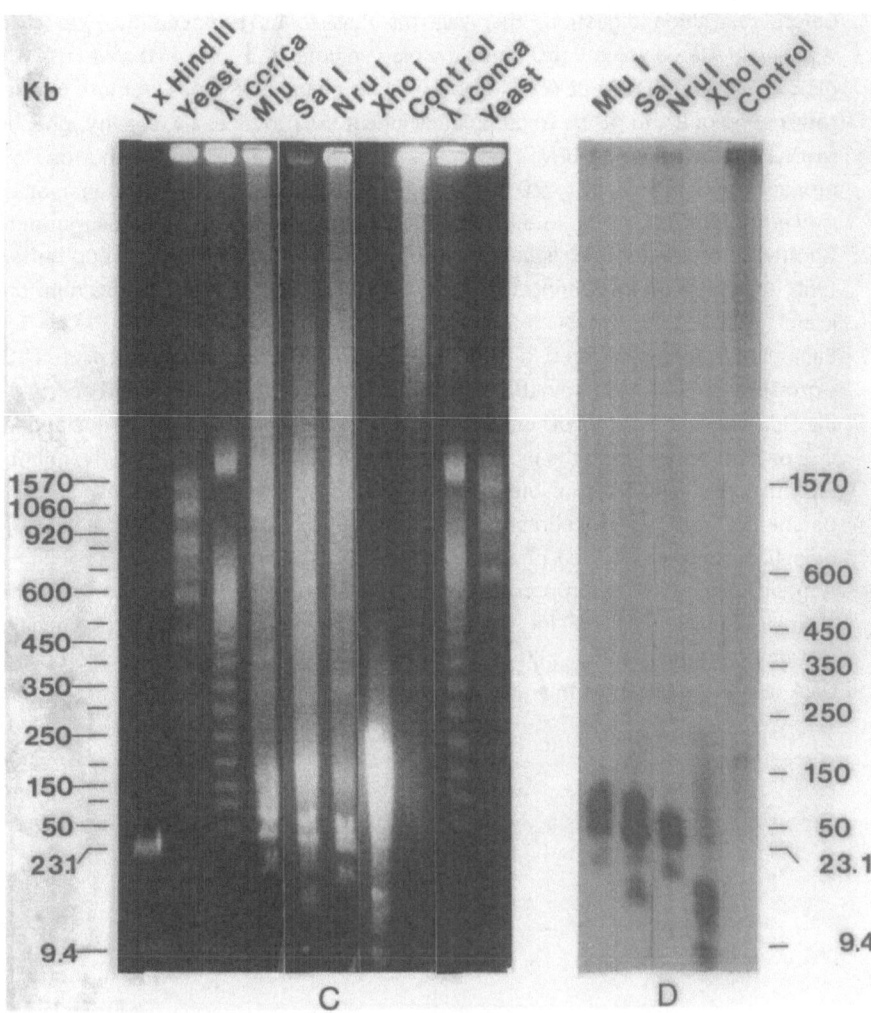

Fig. 3. (*Continued*)

Table 3. Restriction enzymes that generate DNA fragments with an average size of at least 100 kb

Enzyme	Site	Enzyme	Site
Aat II	GACGT/C	*Pvu* I	CGAT/CG
Asu II	TT/CGAA	*Rsr* II	CG/G(A/T)CCG
Bss HII (*Bse* PI)	G/CGCGC	*Sac* II (*Sst* II)	CCGC/GG
Cla I	AT/CGAT	*Sal* I	G/TCGAC
Fsp I (*Mst* I)	TGC/GCA	*Sfi* I	GGCCNNNN/NGGCC
Mlu I	A/CGCGT	*Sma* I	CCC/GGG
Nae I	GCC/GGC	*Sna* BI	TAC/GTA
Nar I	GG/CGCC	*Spl* I	C/GTACG
Not I	GC/GGCCGC	*Xho* I (*Pae* R71)	C/TCGAC
Nru I	TCG/CGA	*Xma* I	C/CCGGG
Pac I	TTAAT/TAA	*Xma* III (*Eag* I)	C/GGCCG

Steps in the procedure

1. Before restriction digestion, the lysis mix has to be removed by extensive washing in a large volume (50 times the plug volume) of $T_{10}E_{10}$ (10 mM Tris-HCl pH 8.0, 10 mM EDTA) at 4 °C. The $T_{10}E_{10}$ is changed 6 times or more over a total period of 24 to 48 h. To completely inhibit the proteinase-K activity, phenyl methyl sulfonyl fluoride (PMSF, prepared as a 0.1 M stock in ethanol or iso-propanol and stored at −20 °C) is added in the first 4 wash steps, at a concentration of 0.1 mM. In this mixture, plugs can be stored for some months.

2. Shortly before restriction digestion, the $T_{10}E_{10}$ is replaced by restriction buffer (only the salts) by incubating the plugs in 1 ml of buffer, twice for 30 min, on ice.

3. Each plug is then incubated in 250 μl digestion buffer supplemented with 100 μg/ml bovine serum albumin (BSA), 1 mM dithiotreitol (DTT) and 8 mM spermidine-HCl. Enzyme (50 to 100 units total; 10 to 20 units per μg of DNA) is added in 2 or 3 portions during the incubation period which lasts from 6 h to overnight. Enzyme activity can be inhibited by sulfonated polyhydrates present in some batches of agarose [H. Lehrach, pers. comm.]. This can be easily checked by digesting some embedded phage lambda DNA.

4. A good indication for proper restriction digestion is to look in the ethidium bromide-stained FIGE gel for the appearance of the chloroplast DNA-specific restriction fragments (Fig. 2). For *Not* I this approach is not usuful as chloroplast DNA is not digested by this enzyme.

Pulsed-field gel electrophoresis

Introduction

Routinely we employ the FIGE system, consisting of a horizontal BRL-electrophoresis gel box (20 cm × 25 cm gels), a switchbox (field inversion unit Mark 4, Biocent) to reverse the electric field and a computer (Olivetti M19) equipped with a program developed by van Ommen (Leiden University, 2333 AL Leiden, The Netherlands) to control the pulse times. Power is supplied by a BioRad unit (model 500/200). Electrophoresis buffer is cooled to approximately 11 °C by circulating it through a reflux cooling spiral which is kept at 4 °C (LKB, 2219 Multitemp II). For more details concerning the FIGE setup we refer to van Daelen *et al.* [113].

Steps in the procedure

1. Prepare the agarose gel (usually 1% (w/v)) in 0.5× TBE (0.045 M Tris, 0.045 M borate and 2 mM EDTA). When pouring the gel, ensure that the tray is perfectly horizontal to prevent the formation of distorted lanes during electrophoresis.
2. Place the gel into the gel box, add the electrophoresis buffer (0.5× TBE) and load the plugs by pushing them into the slots of the gel. It is important that plugs are loaded to the front of the slot and well below the surface of the gel. Also care has to be taken that no air bubbles are present between the plug and the gel; we therefore prefer to load the gel while submerged in the buffer. To avoid loss of the plugs from the slots, they can be sealed by pipetting some low-melting-point agarose on top of them. When slot-sized plugs are applied, however, sealing is not necessary, provided the gel is handled with care in subsequent treatments.
3. Electrophorese under the appropriate conditions (see Table 2). To obtain separation up to 1000 kb, a time ramp of 1 to 60 s, a buffer temperature of 11 °C and a field strength of 7 V/cm should be applied.
4. Stain the gel in 500 to 800 ml of 0.5× TBE to which ethidium bromide is added (1.5 μg/ml) for 1 to 3 h at room temperature. Destain the gel for 3 to 5 h, or overnight, in 0.5× TBE to obtain a clearer picture.

Southern blotting and hybridization

Introduction

The procedures for blotting and subsequent hybridization of pulsed-field gels are similar to conventional gels. A very important step is to degrade large DNA molecules prior to blotting. This is achieved by UV irradiation.

Probes are usually prepared by either nick translation [73] or random primed labelling [41] to a specific activity of 10^8 to 5×10^9 cpm/μg. Hybridization is carried out in rotating bottles, containing 15 ml hybridization mix (consisting of 1 M NaCl, 1% (w/v) SDS, 5% (w/v) dextran sulfate and salmon sperm DNA (1 mg/ml)) for 20 cm $\times$ 25 cm blots. Per ml of hybridization mix 5 ng of probe is added. Hybridization is performed overnight at 65 °C and subsequent washes are at 0.2$\times$ SSC, 1% (w/v) SDS at 65 °C.

An example of a Southern blot analysis is shown in Fig. 2, following hybridization with either a single-copy sequence or total chloroplast DNA.

Steps in the procedure
1. Irradiate the ethidium bromide-stained gel for 5 min, at 302 nm UV light (7 mW/cm^2) to break the DNA molecules. When shorter wavelengths are used, exposure times have to be reduced accordingly (e.g. at 245 nm a 1 min incubation is sufficient). Strand breaking by acid depurination (at 0.25 M HCl) followed by alkaline strand cleavage yields very poor and unreproducible results [111, 113].
2. Denature the DNA by incubating the gel for 30 min in 0.4 M NaOH, 0.6 M NaCl, with gentle agitation.
3. Neutralize the gel (1.5 M NaCl, 0.5 M Tris-HCl 7.5) for 30 min and blot the DNA [73] onto a hybridization membrane, using 10$\times$ SSC (1$\times$ SSC [standard saline citrate]: 0.15 M NaCl, 0.015 M Na$_3$ citrate $\cdot$ 2H$_2$O, pH 7.0) as transfer solution.
4. After blotting, wash the membrane with 0.2 M Tris-HCl, 2$\times$ SSC and dry at room temperature.

High-molecular-weight markers

Introduction

For PFG electrophoresis, the conventional molecular weight markers do not suffice and special high-molecular-weight markers have to be prepared. The most convenient are phage lambda concatemers (for the range from 50 to 1500 kb), yeast chromosomes (for the range from 240 to 2000 kb) and *S. pombe* chromosomes (for the range from 3000 to 6000 kb). However, there are large differences between the chromosome sizes of the various yeast and *S. pombe* strains and only well characterized strains should be used (e.g. AB972 [sizes determined by Link and Olson] or YP148 [sizes determined by Hieter] for *S. cerevisiae* and 975(h$^+$) [39] for *S. pombe*). Organisms which eventually might be used as source of markers for larger molecules are *C. albicans* (for the size range from 1000 to 5000 kb [103]) or *N. crassa,* the chromosomes of which are now estimated to range from 4000 kb to 12,500 kb [85].

Steps in the procedure

Phage lambda concatemers

1. Isolate wild-type phage particles from an infected *Escherichia coli* (e.g. LE392) culture by PEG precipitation [73]. Purified, commercially available, DNA can be used, though, after repeated freezing and thawing, only poor concatemerization is obtained [111, 113]. We therefore prefer to embed intact phage particles, which are subsequently lysed.
2. Resuspend the precipitate in 4 volumes of SM (0.1 M NaCl, 8 mM MgSO$_4$ · 7H$_2$O, 50 mM Tris-HCl (pH 7.5) and 0.01% (w/v) gelatin) and warm to 40 °C.
3. Mix with an equal volume of 1% (w/v) LMP agarose, made in SM and pipet into a pre-chilled plug former.
4. Collect the hardened plugs (approx. 30 min on ice) and incubate in lysis mix for 4 h at 50 °C. The plugs should be completely clear after incubation.
5. Store the plugs at 4 °C in fresh lysis mix. After one day concatemers are formed.

Yeast chromosomes

1. Grow cells overnight in YPD (1% (w/v) yeast extract, 2% (w/v) bacto-trypton, 2% (w/v) dextrose) at 30 °C, under vigorous shaking.
2. Dilute the culture 10 times and allow to grow until an OD$_{600}$ of 1.0 is attained.
3. Collect the cells by centrifugation (3000 *g*, 5 min, 4 °C), wash once with a fifth of the original volume of SE (75 mM NaCl, 25 mM EDTA) and resuspend pellet up in 4 volumes SE.
4. Add DTT (final concentration of 10 mM) and Zymolyase-100T (Kirin Breweries; 6 units/ml), warm the suspension to 40 °C and mix with an equal volume of 1% (w/v) LMP agarose made in SE.
   Other enzymes, such as Lyticase (Sigma), or protoplast-forming enzyme (Boehringer), can be used instead of Zymolyase.

94

5. Prepare plugs as described for plant DNA.
6. Incubate the plugs in an equal volume of SE, DTT, Zymolyase-100T at 37 °C for 2 h.
7. Lyse and store as described for high-molecular-weight plant DNA.

Acknowledgment

We thank Dr. R. Pennell for critically reading the manuscript.

References

1. Albertsen HM, Le Paslier D, Abderrahim H, Dausset J, Cann H, Cohen D (1989) Improved control of partial DNA restriction digestion in agarose using limiting concentrations of Mg^{2+}. Nucleic Acids Res 17: 808.
2. Anand R, Southern EM. Pulsed field gel electrophoresis. In: Rickwood D, Hames BD (eds) Gel Electrophoresis of Nucleic Acids: a Practical Approach. Oxford: IRL Press (In press).
3. Anand R (1986) Pulsed field gel electrophoresis: a technique for fractionating large DNA molecules. Trends Genet 2: 278–283.
4. Antequera F, Bird AP (1988) Unmethylated CpG islands associated with genes in higher plant DNA. EMBO J 7: 2295–2299.
5. Bancroft I, Wolk CP (1988) Pulsed Homogeneous Orthogonal field Gel Electrophoresis (PHOGE). Nucleic Acids Res 16: 7405–7418.
6. Barlow DP, Lehrach H (1987) Genetics by gel electrophoresis: the impact of pulsed field gel electrophoresis on mammalian genetics. Trends Genet 3: 167–171.
7. Bautsch W (1988) Rapid physical mapping of the *Mycoplasma mobile* genome by two-dimensional field inversion gel electrophoresis. Nucleic Acids Res 16: 11461–11467.
8. Beckmann JS, Soller M (1986) Restriction fragment length polymorphism in plant genetic improvement. In: Milfin BJ (ed), Oxford Surveys of Plant Molecular Biology, vol. 3, pp. 196–250. Oxford: Oxford University Press.
9. Beckmann JS, Soller M (1989) Genome genetics in plant breeding. Symposium 2: genome organisation. Vortr Pflanzenzüchtung 16: 91–106.
10. Bernards A, Kooter JM, Michels PAM, Moberts RMP, Borst P (1986) Pulsed field gradient electrophoresis of DNA digested in agarose allows the sizing of the larger duplication unit of a surface antigen in Trypanosomes. Gene 42: 313–322.
11. Beverly SM (1988) Characterization of the 'unusual' mobility of large circular DNAs in pulsed field-gradient electrophoresis. Nucleic Acids Res 16: 925–939.
12. Bird AP (1980) DNA methylation and the frequency of CpG in animal DNA. Nucleic Acids Res 8: 1499–1504.
13. Birren BW, Lai E, Clark SM, Hood L, Simon MI (1988) Optimized conditions for pulsed field gel electrophoresis separations of DNA. Nucleic Acids Res 16: 7563–7582.
14. Birren BW, Lai E, Hood L, Simon MI (1989) Pulsed field gel electrophoresis techniques for separating 1- to 50-kilobase DNA fragments. Anal Biochem 177: 282–286.
15. Birren BW, Simon MI, Lai E (1990) The basis of high resolution separation of small DNAs by asymmetric-voltage field inversion gel electrophoresis and its application to DNA sequencing gels. Nucleic Acids Res 18: 1481–1487.
16. Bostock C (1988) Parameters of field inversion gel electrophoresis for the analysis of pox virus genomes. Nucleic Acids Res 16: 4239–4252.
17. Brown WRA (1988) A physical map of the human pseudoautosomal region. EMBO J 7: 2377–2385.
18. Brown WRA, Bird AP (1986) Long-range restriction site mapping of mammalian genomic DNA. Nature 322: 477–481.
19. Burke DT, Carle GF, Olson MV (1987) Cloning of large segments of exogenous DNA into yeast by means of artificial chromosome vectors. Science 236: 806–812.
20. Burmeister M, Lehrach H (1986) Long-range restriction map around the Duchenne dystrophy gene. Nature 324: 582–585.
21. Cantor CR, Gaal A, Smith C (1988) High resolution separation and accurate size determination in pulsed-field gel electrophoresis of DNA. 3 Effect of electric field shape. Biochemistry 27: 9216–9221.
22. Cantor CR, Smith CL, Mathew MK (1988) Pulsed-field gel electrophoresis of very large DNA molecules. Annu Rev Biophys Biophys Chem 17: 287–304.
23. Carle GF, Frank M, Olson MV (1986) Electrophoretic separation of large DNA molecules by periodic inversion of the electric field. Science 232: 65–68.

24. Carle GF, Olson MV (1984) Separation of chromosomal DNA molecules from yeast by orthogonal-field-alternation-gel electrophoresis. Nucleic Acids Res 12: 5647–5664.

25. Carle GF, Olson MV (1985) An electrophoretic karyotype for yeast. Proc Natl Acad Sci USA 82: 3756–3760.

26. Carle GF, Olson MV (1987) Orthogonal Field Alternation-Gel Electrophoresis. Meth Enzymol 155: 468–482.

27. Chu G, Vollrath D, Davis RW (1986) Separation of large DNA molecules by contour-clamped homogeneous electric fields. Science 234: 1582–1585.

28. Clark SM, Bruce EL, Birren BW, Hood L (1989) A novel instrument for separating large DNA molecules with pulsed homogeneous electric fields. Science 241: 1203–1205.

29. Collins FS (1988) Chromosome jumping. In: Davies KE (ed), Genome analysis: a practical approach, pp. 73–94. Oxford: IRL Press.

30. Collins FS, Drumm ML, Cole JL, Lockwood WK, Van den Woude CF, Ianuzzi MC (1987) Construction of a general human chromosome jumping library, with application to cystic fibrosis. Science 235: 1046–1049.

31. Collins FS, Weissman SM (1984) Directional cloning of DNA fragments at a large distance from an initial probe: a circularization method. Proc Natl Acad Sci USA 81: 6812–6816.

32. Cook PR (1984) A general method for preparing intact nuclear DNA EMBO J 3: 1837–1842.

33. Daniels DL, Olson CL, Brumley R, Blattner FR (1990) Field inversion gel electrophoresis applied to the rapid, multi-enzyme restriction mapping of phage lambda clones. Nucleic Acids Res 18: 1312.

34. Deutsch JM (1989) Explanation of anomalous mobility and birefringence measurements found in pulsed field electrophoresis. J Chem Phys 90: 7436–7441.

35. Deutsch JM, Madden TL (1989) Theoretical studies of DNA during electrophoresis. J Chem Phys 90: 2476–2485.

36. Devos KM (1989) Preparation of plant DNA for separation by pulsed field gel electrophoresis. Electrophoresis 10: 267–268.

37. Ellis THN, Cleary WG, Burcham KWG, Bowen BA (1987) Ramped field inversion gel electrophoresis: a cautionary note. Nucleic Acids Res 15: 5489.

38. Estivill X, Farrall M, Scamber PJ, Bell GM, Hawley KMF, Lench NJ, Bates GP, Kruyer HC, Frederick PA, Stanier P, Watson EK, Williamson R, Wainwright BJ (1987) A candidate for the cystic fibrosis locus isolated by selection for methylation free islands. Nature 326: 840–846.

39. Fan JB, Chikashige Y, Smith CL, Niwa O, Yanagida M, Cantor CR (1988) Construction of a *Not* I restriction map of the fission yeast *Schizosaccharomyces pombe* genome. Nucleic Acids Res 17: 2801–2818.

40. Fangman WL (1978) Separation of very large DNA molecules by gel electrophoresis Nucleic Acids Res 5: 653–665.

41. Feinberg AP, Vogelstein B (1983) A technique for radiolabelling DNA restriction endonuclease fragments to high specific activity. Anal Biochem 132: 6–10.

42. Frearson EM, Power JB, Cooking EC (1973) The isolation, culture and regeneration of *Petunia* leaf protoplasts. Dev Biol 33: 130–137.

43. Galbraith DW, Harkins KR, Maddoz JM, Ayres NM, Sharma DP, Firoozabady E (1983) Rapid flow cytometric analysis of the cell cycle in intact plant tissues. Science 220: 1049–1051.

44. Ganal MW, Tanksley SD (1989) Analysis of tomato DNA by Pulsed Field Gel Electrophoresis. Plant Mol Biol Rep 7: 17–27.

45. Ganal MW, Young ND, Tanksley SD (1989) Pulsed field gel electrophoresis and physical mapping of large DNA fragments in the *Tm-2a* region of chromosome 9 in tomato. Mol Gen Genet 215: 395–400.

46. Gardiner K, Laas W, Patterson D (1986) Fractionation of large mammalian DNA restriction fragments using vertical pulsed-field gradient gel electrophoresis. Som Cell Mol Gen 12: 185–195.

47. Gemmill RM, Coyle-Morris JF, McPeek FD, Ware-Uribe LF, Hecht F (1987) Construction

of long-range restriction maps in human DNA using pulsed field gel electrophoresis. Gene Anal Techn 4: 119–131.

48. Goodfellow PN (1986) Duchenne Muscular Dystrophy: Collaboration and progress. Nature 322: 12–13.

49. Goodfellow PN (1987) Classical and reverse genetics. Nature 326: 824

50. Graham MY, Otani T, Boime I, Olson MV, Carle GF, Chaplin D (1987) Cosmid mapping of the human chorionic gonadotropin β-subunit genes by field inversion gel electrophoresis. Nucleic Acids Res 15: 4437–4448.

51. Gruenbaum Y, Naveh-Maney T, Cedar H, Razin A (1981) Sequence specificity of methylation in higher plant DNA. Nature 292: 860–862.

52. Guzman P, Ecker JR (1988) Development of large DNA methods for plants: molecular cloning of large segments of *Arabidopsis* and carrot DNA into yeast. Nucleic Acids Res 16: 11091–11105.

53. Hanish J, McClelland M (1989) Controlled partial restriction digestions of DNA by competition with modification methyltransferases. Anal Biochem 179: 357–360.

54. Hanish J, McClelland M (1990) Methylase-limited partial *Not*I cleavage for physical mapping of genomic DNA. Nucleic Acids Res 18: 3287–3291.

55. Hardy DA, Bell JI, Long EO, Lindsten T, McDevitt HO (1986) Mapping of the class II region of the major histocompatibility complex by pulsed-field gel electrophoresis. Nature 323: 453–455.

57. Helentjaris TM, Burr B (eds) (1989) Current Communications in Molecular Biology. Development and Application of Molecular Markers to Problems in Plant Genetics. Cold Spring Harbor, NY: Cold Spring Harbor Laboratory.

58. Helentjaris TM, Slocum S, Wright A, Schaefer A, Nienhuis J (1986) Construction of genetic linkage maps in maize and tomato using restriction fragment length polymorphisms. Theor Appl Genet 72: 761–769.

59. Heller C, Pohl FM (1989) A systematic study of field inversion gel electrophoresis. Nucleic Acid Res 17: 5989–6003.

60. Hennekes H, Kuehn S (1989) Control of pulsed field gel electrophoresis at short switching intervals by a microcomputer. Anal Biochem 183: 80–83.

61. Hightower RC, Metge DW, Santi DV (1987) Plasmid migration using orthogonal field-alternation gel electrophoresis. Nucleic Acids Res 15: 8387–8398.

62. Hille J, Koornneef M, Ramanna MS, Zabel P (1989) Tomato: a crop species amenable to improvement by cellular and molecular methods. Euphytica 42: 1–23.

63. Kenwrick S, Patterson M, Speer A, Fischbeck K, Davies K (1987) Molecular analysis of the Duchenne muscular dystrophy region using pulsed field gel electrophoresis. Cell 48: 351–357.

64. Kerem B, Rommens JM, Buchanan JA, Markiewicz D, Cox TK, Chakravarti A, Buchwald M, Tsui LC (1989) Identification of the cystic fibrosis gene: genetic analysis. Science 245: 1073–1080.

65. Klein-Lankhorst R, Rietveld P, Machiels B, Verkerk R, Weide R, Gebhardt C, Koornneef M, Zabel P (1991) RFLP markers linked to the root-knot nematode resistance gene Mi in tomato. Theor Appl Genet 81: 661–667.

66. Klotz LC, Zimm BH (1972) Size of DNA determined by viscoelastic measurements: results on bacteriophages, *Bacillus subtilis* and *Escherichia coli*. J Mol Biol 72: 779–800.

67. Lalande M, Noolandi J, Turmel C, Rousseau J, Slater GW (1987) Pulsed-field electrophoresis: application of a computer model to the separation of large DNA molecules. Proc Natl Acad Sci USA 84: 8011–8015.

68. Landry BS, Michelmore RW (1987) Methods and applications of restriction fragment length polymorphism analysis to plants. In: Bruening G, Harada J, Kosuge T, Hollaenders A (eds), Tailoring Genes for Crop Improvement, pp. 25–44. New York: Plenum.

69. Larkin RH (1976) Purification and viability determinations of plant protoplasts. Planta 128: 213–216.

98

70. Lawrence SK, Smith CL, Srivastava CR, Cantor CR, Weissman SM (1987) Macro organizational analysis of the HLA complex. Science 235: 1387–1390.
71. Lawrence SK, Srivastava R, Rigas B, Chorney MJ, Gillespie GA, Smith CL, Cantor CR, Collins FS, Weissman SM (1986) Molecular approaches to the characterization of megabase regions of DNA: applications to the human major histocompatibility complex. Cold Spring Harbor Symp Quant Biol 51: 123–130.
72. Levene SD, Zimm BH (1987) Separation of open-circular DNA using pulsed-field electrophoresis. Proc Natl Acad Sci USA 84: 4054–4057.
73. Maniatis T, Fritsch EF, Sambrook J (1983) Molecular Cloning: A Laboratory Manual. Cold Spring Harbor, NY: Cold Spring Harbor Laboratory.
74. Mathew MK, Hui CF, Smith CL, Cantor CR (1988) High resolution electrophoresis of DNA. 4. Influence of DNA topology. Biochemistry 27: 9222–9226.
75. Mathew MK, Smith CL, Cantor CR (1988) High resolution separation and accurate size determination in pulsed-field gel electrophoresis of DNA. 1. DNA size standards and the effect of agarose and temperature. Biochemistry 27: 9204–9210.
76. Mathew MK, Smith CL, Cantor CR (1988) High resolution separation and accurate size determination in pulsed-field gel electrophoresis of DNA. 2. Effect of pulse time and electric field strength and implications for models of separation proces. Biochemistry 27: 9210–9216.
77. McClelland M, Kessler LG, Bittner M (1984) Site-specific cleavage of DNA at 8- and 10-base-pair sequences. Proc Natl Acad Sci USA 81: 983–987.
78. McClelland M (1983) The frequency and distribution of methylatable DNA sequences in leguminous plant protein coding genes. J Mol Evol 19: 346–354.
79. McPeek FD, Coyle-Morris JF, Gemmill RM (1986) Separation of large DNA molecules by modified pulsed field gradient gel electrophoresis. Anal Biochem 156: 274–285.
80. Meese E, Meltzer P, Trent J (1989) Application of natural partial digests to pulsed-field gel analysis of the amplified *MDR* locus. Genomics 5: 371–374.
81. Michiels F, Burmeister M, Lehrach H (1987) Derivation of clones close to *met* by preparative field inversion gel electrophoresis. Science 236: 1305–1308.
82. Nasmyth K, Sulston J (1987) High-altitude walking with YAC's. Nature 328: 380–381.
83. Nelson M, Christ C, Schildkraut I (1984) Alteration of apparent restriction endonuclease specificities by DNA methylation. Nucleic Acids Res 12: 5165–5173.
84. Olschwang S, Thomas G (1989) Temperature gradient increases FIGE resolution. Nucleic Acids Res 17: 2363.
85. Orbach MJ, Vollrath D, Davis RW, Yanofsky C (1988) An electrophoretic karyotype of *Neurospora crassa*. Mol Cell Biol 8: 1469–1473.
86. Orkin SH (1986) Reverse genetics and human disease. Cell 47: 845–850.
87. Overhauser J, Radic MZ (1987) Encapsulation of cells in agarose beads for use with pulsed-field gel electrophoresis. Focus 9: 8–9.
88. Pohl TM, Zimmer M, MacDonald ME, Smith B, Bucan M, Poustka A, Volinia S, Searle S, Zehetner G, Wasmuth JJ, Gusella J, Lehrach H, Frischauf AM (1988) Construction of a *Not*I linking library and isolation of new markers close to Huntington's disease gene. Nucleic Acids Res 16: 9185–9198.
89. Poustka A, Lehrach H (1986) Jumping libraries and linking libraries: the next generation of molecular tools in mammalian genetics. Trends Genet 2: 174–179.
90. Poustka A, Lehrach H (1988) Chromosome jumping: a long range cloning technique. In: Setlow JM (ed), Genetic Engineering, Vol 10, Plenum, pp. 169–195. New York.
91. Rappold G, Lehrach H (1988) A long range restriction map of the pseudoautosomal region by partial digest PFGE analysis from the telomere. Nucleic Acids Res 16: 5361–5377.
92. Riordan JR, Rommens JM, Kerem B, Alon N, Rozamhel R, Grzelczak Z, Zielenski J, Lok S, Plavsic N, Chou JL, Drumm ML, Ianuzzi C, Collins FS, Tsui LC (1989) Identification of the cystic fibrosis gene: cloning and characterization of complementary DNA. Science 245: 1066–1073.
93. Rommens JM, Ianuzzi MC, Kerem B, Drumm ML, Melmer G, Dean M, Rozmahel R, Cole

JL, Kennedy D, Hidaka N, Zsiga M, Buchwald M, Riordan JR, Tsui LC, Collins FS (1989) Identification of the cystic fibrosis gene: chromosome walking and jumping. Science 245: 1059–1065.

94. Sarfatti M, Katan J, Fluhr R, Zamir D (1989) An RFLP marker in tomato linked to the *Fusarium oxysporum* resistance gene *I2*. Theor Appl Genet 78: 755–759.

95. Schwartz DC, Cantor CR (1984) Separation of yeast chromosome-sized DNA by pulsed field gradient gel electrophoresis. Cell 37: 67–75.

96. Schwartz DC, Koval M (1989) Conformational dynamics of individual DNA molecules during gel electrophoresis. Nature 338: 520–522.

97. Schwartz DC, Saffran W, Welsh J, Haas R, Goldenberg M, Cantor CR (1982) New techniques for purifying large DNAs and studying their properties and packaging. Cold Spring Harbor Symp Quant Biol 47: 189–195.

98. Schwartz DC, Smith LC, Baker M, Hsu M (1989) ED: pulsed electrophoresis instrument. Nature 342: 575–576.

99. Simske JS, Scherer S (1989) Pulsed field gel electrophoresis of circular DNA. Nucleic Acids Res 17: 4359–4365.

100. Slater GW, Noolandi J (1988) Electric field gradients and band sharpening in DNA gel electrophoresis. Electrophoresis 9: 643–646.

101. Smith CL, Cantor CR (1987) Purification, specific fragmentation, and separation of large DNA molecules. Meth Enzymol 155: 449–467.

102. Smith CL, Klco SR, Cantor CR (1988) Pulsed-field gel electrophoresis and the technology of large DNA molecules. In: Davies KE (ed), Genome Analysis, A Practical Approach, pp. 41–72. Oxford: IRL Press.

103. Smith SB, Aldridge PK, Callis JB (1989) Observation of individual DNA molecules undergoing gel electrophoresis. Science 243: 203–206.

104. Snell RG, Wilkins RJ (1986) Separation of chromosomal DNA molecules from *C. albicans* by pulsed field gel electrophoresis. Nucleic Acids Res 14: 4401–4406.

105. Sobral BWS, Atherly AG (1989) Pulse time and agarose concentration affect the electrophoretic mobility of cccDNA during electrophoresis in CHEF and FIGE. Nucleic Acids Res 17: 7359–7369.

106. Southern EM, Anand R, Brown WRE, Fletcher DS (1987) A model for the separation of large DNA molecules by crossed field gel electrophoresis. Nucleic Acids Res 15: 5925–5943.

107. Steinmetz M, Minard K, Horvath S, McNicholas J, Srelinger J, Wake C, Lange E, Mach B, Hood L (1982) A molecular map of the immune response region from the major histocompatibility complex of the mouse. Nature 300: 35–42.

108. Tanksley SD, Young ND, Paterson AH, Bonierbale MW (1989) RFLP mapping in plant breeding: new tools for old science. Bio/technology 7: 257–264.

109. Turmel C, Lalande M (1988) Resolution of *Schizosaccharomyces pombe* chromosomes by field inversion gel electrophoresis. Nucleic Acids Res 16: 4727.

110. Upcroft JA, Boreham PFL, Upcroft P (1989) Different grades of agarose affect electrophoretic migration of large DNA molecules. Nucleic Acids Res 17: 3315.

111. Van Ommen GJB, Verkerk JMH (1986) Restriction analysis of chromosomal DNA in a size range up to two million base pairs by pulsed field gradient electrophoresis. In: Davis KE (Ed) Human Genetic Disease, A Practical Approach, pp. 113–133. Oxford: IRL Press.

112. Van Ommen GJB, Verkerk JMH, Hofker MH, Monaco AP, Kunkel LM, Ray R, Worton R, Wieringa B, Bakker E, Pearson PL (1986) A physical map of 4 milion bp around the Duchenne muscular dystrophy gene on the human X-chromosome. Cell 47: 499–504.

113. Van Daelen RAJ, Jonkers JJ, Zabel P (1989) Preparation of megabase-sized tomato DNA and separation of large restriction fragments by field inversion gel electrophoresis (FIGE). Plant Mol Biol 12: 341–352.

114. Vollrath D, Davis RW (1987) Resolution of DNA molecules greater than 5 megabases by contour-clamped homogeneous electric fields. Nucleic Acids Res 15: 7865–7876.

115. Ward ER, Jen GC (1990) Isolation of single-copy-sequence clones from a yeast artificial

chromosome library of randomly-sheared *Arabidopsis thaliana* DNA. Plant Mol Biol 14: 561–568.

116. West R (1987) The electrophoretic mobility of DNA in agarose gels as function of temperature. Biopolymers 26: 607–608.

117. Widholm LM (1972) The use of fluorescein diacetate and phenosafranine for determining viability of cultured plant cells. Stain Technol 47: 189–194.

118. Wieringa B, Bakker E, Pearson PL (1980) A physical map of 4 million bp around the Duchenne muscular dystrophy gene in the human X chromosome. Cell 47: 499–504.

119. Young ND, Zamir D, Ganal MW, Tanksley SD (1988) Use of isogenic lines and simultaneous probing to identify DNA markers tightly linked to the Tm-2a gene in tomato. Genetics 120: 579–585.

6. Physical mapping by random clone fingerprint analysis

BRIAN M. HAUGE & HOWARD M. GOODMAN

Department of Genetics, Harvard Medical School and Department of Molecular Biology, Massachusetts General Hospital Boston, MA 02114, USA

Introduction

One of the major challenges of molecular biology is the isolation of genes where the biochemical function of the gene product is unknown. As described in the preceding chapters, an increasing array of techniques are available for identifying DNA probes residing within one to several cM of the gene of interest. To clone the gene it is then necessary to bridge the intervening gap, often spanning distances of millions of base pairs which typically separate genetically linked markers in both mammals and higher plants. While mapping such large regions is theoretically possible with the traditional technique of 'chromosome walking' using overlapping cosmid or λ clones, in practice the procedure is extremely labor-intensive and ill-suited for large projects where more than a few steps are required.

Recently, interest has focused on strategies for constructing physical maps of entire genomes. By definition, a physical map consists of a linearly ordered set of DNA fragments encompassing the genome or region of interest. Physical maps are of two types, macro-restriction maps and ordered clone maps. The former consists of an ordered set of large DNA fragments generated by using restriction enzymes whose recognition sequences are infrequently represented in the genome ([60]; see also van Daelen and Zabel, this book). The macro-restriction map provides information about the organization of DNA fragments at the level of the intact chromosome, thereby providing long-range continuity. As the name implies, an ordered clone map consists of an overlapping collection of cloned DNA fragments. The DNA may be cloned into any one of the available vector systems: yeast artificial chromosomes (YACs), cosmids, phage or even plasmids. The major advantages of ordered clone maps is that they are of high resolution and they directly provide the clones for further study. In all likelihood, complete physical mapping of complex genomes will require a combination of techniques.

The benefits of having a physical map are two-fold. First, the map provides access to any region of the genome which can be genetically identified. In other words, the physical map serves as a cloning tool by facilitating the movement from the genetic locus to the cloned gene. Given a mutation of known genetic map location, the physical map can be used to easily isolate an overlapping collection of clones encompassing the locus of interest. By eliminating the need

101

J.S. Beckmann and T.C. Osborn (eds) Plant Genomes: Methods for Genetic and Physical Mapping, 101–139.
© 1992 *Kluwer Academic Publishers.*

for labor-intensive steps such as chromosome walking, researchers are free to focus their efforts on the isolation and characterization of the gene of interest. Second, physical maps provide a starting point for studying global genomic organization. As an increasing number of genes are cloned and molecular biological information is accumulated, one can begin to investigate the physical linkage of cloned genes, study the organization and distribution of repetitive elements and address questions such as how physical distance and genetic distance are correlated. In this context, the map provides the framework for cataloging and integrating molecular biological information. Ultimately, genome organization will be investigated at the nucleotide level. Clearly, physical maps are the logical substrates for genome sequencing projects.

We are engaged in a project to construct a complete physical map of the *Arabidopsis thaliana* genome which will ultimately consist of a fully overlapping collection of cloned DNAs encompassing the five *Arabidopsis* linkage groups [23]. There are several reasons for using *Arabidopsis* as a model system for the study of plant biology. Its short life cycle, small size and large seed output make it well suited for classical genetic analysis [38]. Mutations have been described affecting a wide range of fundamental developmental and metabolic processes [16, 38]. A genetic linkage map consisting of some 100 loci has been assembled [29] and an increasing number of cloned genes and restriction fragment length polymorphisms (RFLPs) are available for correlation of the genetic map with the physical map [10, 40]. For molecular biological studies, *Arabidopsis* offers the additional advantages of having a very small genome (70,000 kb) and a remarkably low content of interspersed repetitive DNA [34, 48]. Both of these features are highly unusual among higher plants and suggest that the *Arabidopsis* genome is ideally suited for a physical mapping project.

In this chapter, we describe the methodologies which we are employing in our effort to construct a physical map of the *Arabidopsis thaliana* genome [23]. The first stage of this project involves a 'fingerprinting' procedure for the identification of overlaps between clones. Clones are picked at random, fingerprinted and assembled into overlapping sets referred to as contigs. In the second stage, clones are selected by hybridization using probes from the ends of contigs, unattached clones and yeast artificial chromosome (YAC) libraries to fill in the gaps.

Fingerprinting

Several efforts have been initiated to assemble physical maps based on the fingerprinting of random clones [13, 23, 28, 45]. The fingerprints are generated by digesting randomly selected clones from primary libraries with one to several restriction enzymes. Following size fractionation by gel electrophoresis (either agarose or polyacrylamide), the lengths of the fragments are determined. The number and the size of the fragments constitute a unique signature or fingerprint of the cloned insert. For fingerprinting, it is unnecessary to generate a restriction map of the clone. The bands must however be descriptive of the insert and the informational content of the fingerprint must be sufficient to make a reliable assignment of overlapping regions. Clones are said to be overlapping when the fingerprints of two clones are sufficiently similar.

The fingerprinting protocols which we describe are based on the method-ologies of Coulson *et al.* [13]. The method for fingerprinting clones is illustrated in Fig. 1. Cloned DNAs are digested with a restriction enzyme having a 6 bp

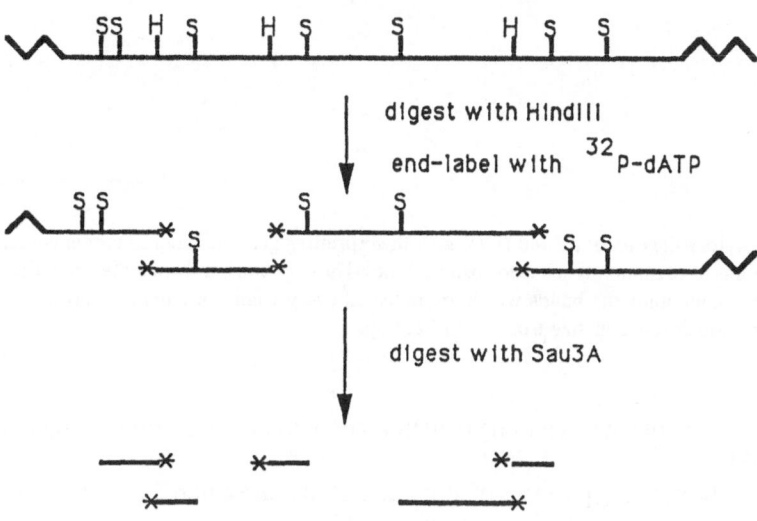

Fig. 1. Fingerprinting procedure. The solid lines depict the cloned insert while the vector is represented by the zig-zag line. ddGTP, dideoxy GTP; H, *Hind* III recognition sequence; S, *Sau*3A recognition sequence. The stars at the ends of the lines represent the end-labeled digestion products.

specificity which leaves staggered ends which are simultaneously labeled with reverse transcriptase and the appropriate nucleoside triphosphates. The reactions are terminated by high temperature and the fragments are subjected to a second round of cleavage with a restriction enzyme having a 4 bp specificity. The resultant fragments are size-fractionated on a denaturing 4% polyacrylamide gel.

Figure 2 shows an example of a typical gel containing the fingerprints of 48 clones with marker lanes interspersed every seven lanes. The markers are a total *Sau*3A digest of lambda DNA (end-labeled) and are used for calibration of the film. For subsequent analysis the position of the bands are entered into

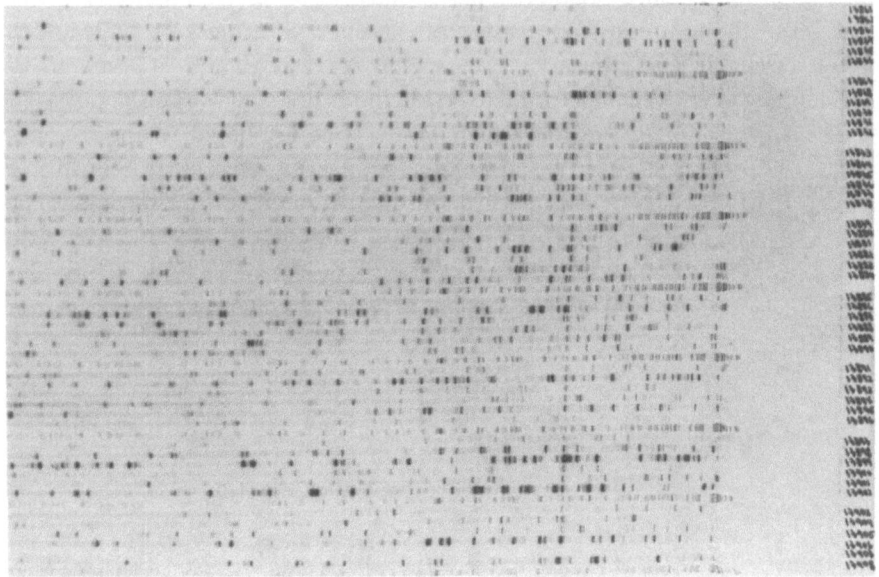

Fig. 2. Autoradiogram of a *Hin*d III/*Sau*3A fingerprinting gel. The autoradiogram contains the fingerprints of 48 randomly selected cosmid clones. The common band originates from the vector. The lanes containing the bands which are repeated every 7 lanes are the end-labeled λ *Sau*3A markers, which range in size from 58 to 2225 bp.

the computer using a scanning densitometer and an image-processing package [62, 63].

Once the banding patterns of individual clones are entered into the computer, they are then compared in a pairwise fashion against the entire data set. The output is a ranked order of the most probable matches (Fig. 3). Based on these numbers, the regions of probable overlap are determined and the clones are assembled into contigs (Fig. 4). It should be noted that the computer does not actually assemble the contigs. Assembly is performed interactively with a computer program [13]. Before the clones are joined, the reliability of the match is assessed by visually aligning the films and the overlap must be logically consistent.

The combination of enzymes, *Hin*d III and *Sau*3A, has been suitable for the *Arabidopsis* project, yielding 25 bands on the average which allows detection of overlaps in the range of 35 to 50% (see Lander and Waterman [32] for statistical considerations). There are two reasons why we chose this combination of enzymes. Many of the cloned *Arabidopsis* genes, as well as some of the RFLPs [10], were isolated from λ libraries which were constructed by either *Sau*3A or *Mbo*I partial digestion. *Sau*3A was therefore used to maintain the same cleavage specificity that was used to make the libraries. In doing so, anomalous bands arising from the insert/vector junction are avoided. In practice, this is not important for cosmid clones where the bands are numerous.

A	B	C		D	E		F	G	H
334-14615	(27b, 0)								
		18	matches	261-11245	(26b,	427)	0.3E-16	4	9 d
		18	matches	112-4348	(28b,	427)	0.1E-15	11	6
		12	matches	262-11273	(15b,	427)	0.3E-12	4	0
		13	matches	284-12345	(22b,	427)	0.2E-10	0	2
		16	matches	409-181202	(28b,	0)	0.1E-12	0	6
	difmap:	13	matches	148-6014	(31b,	427)	0.2E-08	9	6
334-14616	(21b, 0)								
		18	matches	342-14999	(28b,	0)	0.6E-19	0	3 d
		17	matches	272-11743	(32b,	345)	0.4E-16	4	3
		16	matches	264-11353	(26b,	345)	0.8E-16	4	3
		15	matches	395-17555	(37b,	0)	0.5E-12	0	2
		14	matches	335-14680	(29b,	0)	0.6E-12	0	2
	difmap:	5	matches	126-5045	(18b,	496)	0.8E-03	19	3
	difmap:	4	matches	299-13074	(20b,	882)	0.1E-01	0	3
	difmap:	4	matches	374-16506	(21b,	0)	0.1E-01	0	3
	difmap:	12	matches	370-16330	(24b,	0)	0.4E-10	0	3
334-14617	(20b, 0)								
		15	matches	271-11728	(33b,	53)	0.3E-13	2	5 d
		14	matches	164-6802	(44b,	53)	0.5E-10	0	4
		13	matches	235-10263	(36b,	53)	0.9E-10	0	2
		10	matches	11-1234	(29b,	53)	0.6E-07	11	1
		11	matches	328-14425	(36b,	0)	0.3E-07	0	1
	difmap:	4	matches	318-13958	(21b,	0)	0.1E-01	0	4
	difmap:	5	matches	447-2488h	(28b,	431)	0.5E-02	0	4
	difmap:	7	matches	303-13231	(36b,	0)	0.4E-03	0	4

Fig. 3. Output of the clone matching programs. Columns A and B describe the clone being analyzed. The format is as follows: (A) gel number-clone name; (B) the total number of bands used to fingerprint the clone and its contig number. The subsequent columns, C-H, describe possible overlapping clones ranked in decreasing significance according to the following format: (C) the number of bands common to the clone being analyzed and the respective matching clone; (D) gel number-clone name of the matching clone; (E) the total number of bands used to fingerprint the matching clone and its contig number; (F) the probability that the match is due to chance; (G) the distance in bands to the nearest end of the contig; (H) the first line indicates the number of bands from the incoming clone which are not shared by the matching clone, while the following lines describe the number of difference bands common to the subsequent matching clones. The clone matching programs are described in detail by Sulston *et al.* [62].

This is however a consideration when working with λ clones which contain smaller inserts. Second, during the early stages of the project data entry was performed manually. The number of fragments which are generated by labeling the *Hin*d III sites followed by *Sau*3A digestion is manageable for manual data input. With the availability of semi-automatic data entry numerous alternative strategies are possible. For example, overlaps may be established using a restriction enzyme(s) which generates a large number of fragments for each clone. One of the main considerations for choosing an enzyme or combination of enzymes is that the number of fragments generated is optimal for the statistical detection of overlapping regions.

There are two clear limitations of mapping strategies which are based on simple fingerprint analysis. First, the amount of information provided by a

106

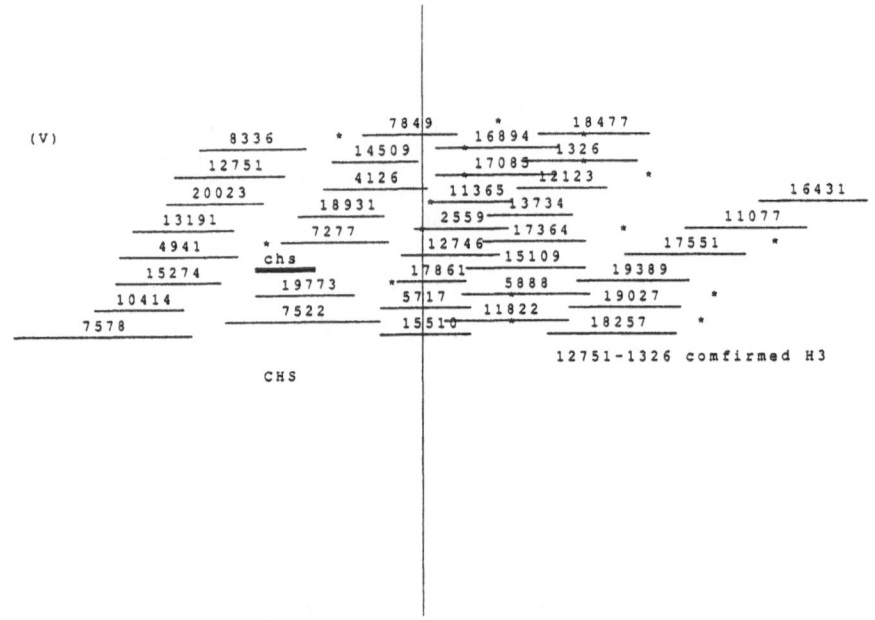

Fig. 4. Computer display of a contig. Clones are depicted by the solid lines, the length of which is proportional to the number of bands in its fingerprint. The length of the lines do not necessarily reflect the size of the clone since the number of bands in the fingerprint is dictated solely by the number of end-labeled recognition sequences. Therefore, clones containing a small number of the appropriate restriction sites appear smaller than clones containing a large number of sites. The asterisk denotes buried clones which are not displayed. In other words, redundant clones are buried to avoid screen clutter. The V in the upper left-hand corner indicates the chromosome number, while comments are displayed below the contig. Contig assembly is as described by Sulston *et al.* [62].

fingerprint is limited; so, in general, rather large overlaps are required to reliably declare a match. The combination of enzymes, *Hin*d III and *Sau*3A, used in the *Arabidopsis* project only permit the detection of overlaps in the range of 35–50%. So smaller overlaps will be ignored in pairwise comparisons. Second, clones containing few or no restriction sites for the enzyme(s) used for generating the labeled ends will be unattached since a minimal number of bands are required for the statistical detection of overlapping regions. Both of these problems can be circumvented by fingerprinting the clones residing at the end of the contigs and the unattached clones using different combinations of enzymes. In doing so, joins which previously went undetected due to statistical limitations can be established since the match probability is now the product of the individual probabilities. In addition, by using several combinations of enzymes it is unlikely that a given clone will have a non-random distribution of restriction sites for all of the enzymes. Consequently, joins can be established for the clones which were previously unattached as a result of having a minimal number of recognition sequences for the enzyme initially employed.

Clearly, two or more enzyme combinations for each clone can be used from the onset of the project. By increasing the amount of information obtained from each clone, the rate of progress is greatly increased. A mathematical analysis of random clone fingerprinting by Lander and Waterman [32] shows that decreasing the minimal detectable overlap from 50 to 25% significantly speeds the progress of a project. On the other hand, relatively little is gained by further reducing the minimal detectable overlap from 25% to the theoretical limit of 0%. Based on this analysis, the authors recommend the use of fingerprinting strategies which detect overlaps in the range of 15 to 20%. While the advantage of using a more sensitive fingerprinting strategy is obvious, it is important to consider the increased effort which is required to obtain a more sensitive fingerprint.

In general, 8 to 10 genomic equivalents must be fingerprinted to achieve between 70 and 90% coverage of the genome. Therefore, the mapping of large genomes is both laborious and computationally intensive. On the other hand, the repetitive nature of the methodologies make the procedures readily amenable to automation. The automation of both template preparation and the subsequent manipulations and enzymatic reactions are quickly becoming a reality. In addition, fluorescence-based approaches which use automatic data collection have recently been described [5, 9] using commercially available automated DNA sequencers [61].

Random fingerprinting procedures are not expected to produce complete physical maps. Instead, the map will consist of many contigs composed of two or more overlapping clones. As the project progresses, the number of contigs decreases as the gaps are closed. After this point, the rate of finding new joins becomes unacceptably low due to the scarcity of the remaining clones. Completion of the map then requires a directed approach since a prohibitively large number of clones would be required to close all of the gaps by random clone fingerprinting.

In addition to the statistical limitations, both the number and the size of the contigs generated by random clone mapping will be strongly influenced by any cloning biases which are encountered. At least two factors contribute to cloning bias: the inability to clone certain regions of the genome using a given host/vector system results in non-representative libraries and non-uniform growth of individual clones leading to sampling bias. To circumvent such problems, it is likely that multiple libraries and multiple host vector systems will be required.

Since the success in completing a mapping project is strongly dependent on the representation of the library(s), it is important to consider the factors which are known to cause libraries to deviate from randomness. There are several types of sequences which have been shown to be under-represented in genomic libraries. It is well established that direct repeats are prone to rearrangement in Rec^+ hosts; under-representation of such sequences can be avoided by using a recA host [3, 59]. In addition, palindromic sequences have been to shown to be unstable or completely non-viable in both λ [33] and in plasmids [35] which are grown in wild-type hosts. Palindromes can however be efficiently propagated in *E. coli* hosts harboring mutations in *rec*B and/or *rec*C and *sbc*B [67]. It should be noted that *Col* E1 origins of replication are poorly maintained in *rec*B, *rec*C, *sbc*B hosts. Vectors based on *Col* E1 can be propagated with hosts harboring an additional mutation in either the *rec*A or the *rec*F genes [2] which further serves to stabilize direct repeats. Finally, there is a report of a 360 bp DNA sequence contained within the *Physarum* actin gene (*ard*A) which suppresses plaque formation of λ clones and causes *rec*A-independent deletions [41]. Interestingly, the sequence which consists of five (dA)(dT) and (dG)(dC) homopolymers can only be cloned in *rec*B, *rec*C, *sbc*B hosts.

Cloning bias may further be introduced by the host restriction systems. Susceptibility to restriction can result from either the presence or absence of base modification of the DNA [53]. One of the best studied examples is the *Eco*K system which restricts DNA when its recognition sequence is unmodified. The majority of hosts which are commonly used for cloning experiments harbor a mutation in the *hsd*R gene which encodes the *Eco*K restriction endonuclease. It is surprising, on the other hand, that most of the published *in vitro* packaging systems are based on extracts containing functional *Eco*K restriction endonuclease [25, 26, 54]. The exceptions are the single-strain *in vitro* packaging system derived from *E. coli* C [54–56] and commercially available packaging extracts (Stratagene). Since *Eco*K restriction works *in vitro* [37], activity present in the extracts has the potential to bias libraries against inserts containing the heptomeric *Eco*K recognition site. It has been reported that *hsd*R activity can account for a 3 to 5 fold bias during library construction. *E. coli* also restricts DNA containing 5-methylcytosine. This restriction is due to two genetically distinct systems, *Mcr*A and *Mcr*B, which differ in their sequence specificity [50, 52]. Several lines of evidence have been presented which suggest that the *Mcr*B system causes a significant bias in mouse genomic libraries [51]. It is expected that methylcytosine-containing DNA from other organisms will be similarly restricted. This is an important consideration when constructing plant genomic libraries since plant DNA is heavily methylated at cytosine residues [57]. In some species 5-methylcytosine accounts for up to 30% of the total cytosines [21].

An exhaustive review of the factors known to cause libraries to deviate from randomness is clearly beyond the scope of this chapter. The choice of vectors, method of library construction and additional host factors which we have not discussed also have the potential to bias libraries. Furthermore, it is likely that unknown factors operate which prevent the propagation of certain sequences. This brief discussion is primarily intended to illustrate some of the potential problems which can be encountered when constructing genomic libraries and to emphasize the importance of making a judicious choice before selecting a host/vector system. Even small cloning biases can compromise both the rate and the final level of completion of mapping projects which are based on random clone analysis.

Contig joining

Once the practical limit of random clone mapping is reached, success in completing a map depends largely on the ability to bridge the remaining gaps. The most viable option is to select the missing clones by hybridization. One approach for selecting linking clones is to make end-probes from unattached clones and clones residing at the end of the contigs. For this approach to be practical it is important to generate end-probes with minimal effort. The cosmid libraries should therefore be constructed in vectors containing convergent bacteriophage promoters (for example Sp6 and T7 promoters) flanking the insert. The end-clones and the unattached clones are picked into microtiter dishes and plated out onto nylon filters in ordered arrays. By probing the cosmid grids with mixed RNA probes (prepared from rows of clones) [17], overlaps which were not detected by fingerprint analysis can be established (see Fig. 5). The use of mixed end-probes is important when a large number of joins must be established since the number of hybridizations required is reduced by a factor of N, where N is the number of clones used to make the probes.

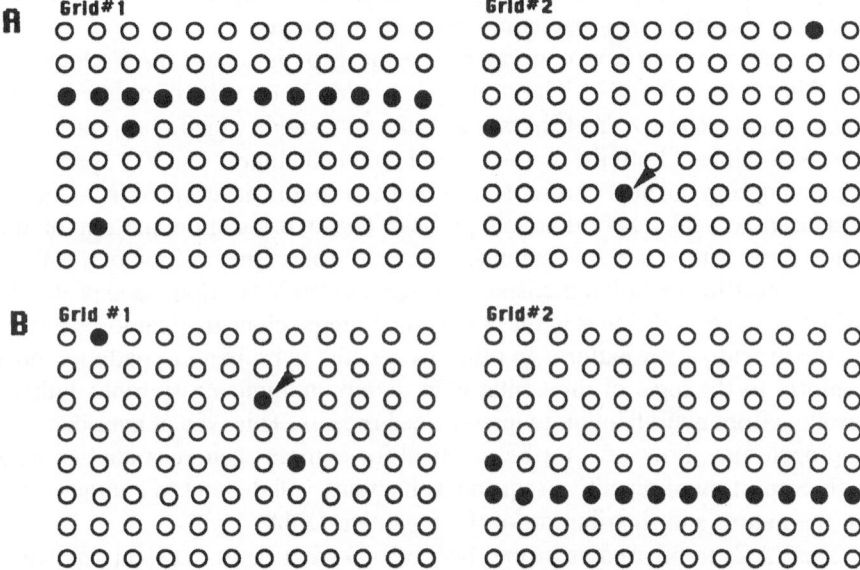

Fig. 5. Strategy for linking contigs by hybridization using mixed end-probes. Cosmid clones from the ends of the contigs and a representative collection of unattached clones are gridded onto nylon membranes from glycerol stocks which are held at − 80 °C in 96-well microtiter plates. Individual clones are assigned a unique position on the grids. RNA probes are prepared from mixed pools of 12 cosmid clones corresponding to a row of the microtiter plate. Hybridization to grids 1 and 2 is shown using as the probe (A) clones pooled from row three of grid number 1 and (B) clones pooled from row six of grid 2. Hybridization to the grids detects both the template clones and the overlapping clones. The procedure is repeated until probes have been prepared from all of the rows. Hybridizing clones which appear in both rows indicate a potential overlap which is indicated by the arrows (panels A and B).

It must be kept in mind that the majority of the gaps are real and that the missing clones are either rare or non-existent in the cosmid libraries which were used for the random clone mapping. Therefore, the end-probes are also used to probe additional libraries based on different host/vector systems. The use of different host/vector systems is intended to eliminate, or at least reduce, cloning bias. In particular, the hybridization to yeast artificial chromosome (YAC) clones is an important component for this analysis.

One of the most important new technical advances in molecular biology is the cloning of megabase-size DNA fragments using YAC vectors [7]. The construction of YAC libraries involves the ligation of large DNA fragments (50–1000 kb) into a vector containing selectable markers and the functional components of a eukaryotic chromosome; ARS elements required for autonomous replication, the centromere which results in proper disjunction during meiosis and mitosis, and telomeres required for the replication of linear molecules [39]. The constructs are transformed into *Saccharomyces cerevisiae* where they are replicated along with the endogenous chromosomes. There are two clear advantages of using the yeast cloning system. The large size of YAC clones means that fewer clones must be examined. Equally as important, YACs offer the potential to give a random or at least different representation of clones than are obtained using bacterial host/vector systems.

A complementary approach to bridge the gaps is to use YAC clones as hybridization probes [14]. The strategy is to prepare two sets of ordered grids: one of a representative YAC library and one of cosmids which is as representative as possible of both the contigs and unattached clones. The YACs are then separated from the host chromosomes by electrophoresis, isolated from the gel and used to make hybridization probes. The hybridization pattern of the cosmid grid is then used to establish linkage as well as the position of the YAC with respect to the ordered cosmids. Since a given YAC clone is expected to hybridize to several clones in the contig, the ordered clone map imposes a logic on the hybridization pattern. In other words, the hybridization patterns must conform to the logic of the contig map thereby minimizing spurious linkage resulting from hybridization to interspersed repeats. This type of hybridization approach may, however, not be practical for complex genomes containing a high proportion of repetitive sequences; its utility will depend to a large extent on the nature and distribution of the repetitive DNA.

The YACs to be used as probes may be picked at random or, alternatively, selected from the YAC grid based on hybridization with cosmids as described above. One requirement of this approach is that the cosmid vectors have no significant homology to the YAC vectors. This permits the direct hybridization of the YACs to cosmids and *vice versa* thereby eliminating the need to first separate the insert from the vector sequences. The Lorist [15] series of cosmid vectors have been successfully used for this approach [14].

In the future, it is likely that YAC clones will be used at the onset of physical mapping projects. Using existing technology it is possible to fingerprint YACs directly [31]. Moreover, the ability to easily generate end-probes from YACs using techniques such as inverse PCR [44] allows for the construction of physical maps based on hybridization strategies. The application of mapping strategies based on YACs should make it possible to undertake projects which are an order of magnitude larger than those currently underway. It is unlikely, however, that YACs will supersede cosmid and λ clone maps since the smaller clones are generally required for routine procedures such as DNA sequencing and gene isolation. Whether it is more efficient to assemble a YAC map and subsequently isolate the smaller clones when needed or use YACs to link a cosmid map remains to be determined.

Procedures

Isolation of genomic DNA

The two procedures given here have been used for the isolation of DNA from *Arabidopsis*, but are generally applicable to other plants. The first procedure describes the extraction of DNA from nuclei which is used to eliminate, or at least reduce, the presence of undesirable plastid DNA. The second procedure describes the isolation of total DNA which is both simple and gives good yields of DNA. However, DNA prepared by this procedure is not recommended for the construction of libraries used for the analysis of randomly selected clones since the presence of clones containing inserts of plastid DNA will increase the total number of clones which must be examined to construct a physical map. For example, in *Arabidopsis* plastid DNA accounts for up to 20% of total DNA [48]. Libraries constructed from total DNA are, however, suitable when clones are selected by hybridization.

Steps in the procedure

Isolation of nuclear DNA (based on the method of Hamilton *et al.* [22])
1. Harvest 100 g of tissue which has been destarched by placing the plants in the dark for 48 h.

Note
1 All subsequent steps are performed at 4 °C unless indicated differently.

2. Wash with ice-cold water and cut into small pieces using a single-edge razor blade.
3. Cover the tissue with ice-cold diethyl ether and stir for 3 min. Decant the ether and rinse well with ice-cold water to remove the residual ether.
4. Add 300 ml of buffer A (1 M sucrose, 10 mM Tris-HCl pH 7.2, 5 mM MgCl$_2$, 5 mM β-mercaptoethanol and 400 μg/ml ethidium bromide). The inclusion of ethidium bromide is essential for the isolation of high-molecular-weight DNA. Homogenize tissue with either a polytron or Waring blender at medium speed for 1–3 min.
5. Filter the homogenate through 4 layers of cheese cloth, then through 2 layers of Miracloth (Calbiochem).
6. Centrifuge at 9000 rpm in a Beckman JA-10 or equivalent rotor for 15 min.
7. Decant and discard the supernatant and resuspend the pellet in 50 ml of buffer A plus 0.5% Triton X-100 using a homogenizer with a teflon pestle. Transfer to two 30 ml Corex tubes and centrifuge at 8000 rpm for 10 min in a Beckman JS-13 rotor.
8. Repeat step 7, except centrifugation is at 6000 rpm for 10 min.
9. Resuspend the pellet in 10 ml of buffer A plus 0.5% Triton X-100. Layer crude nuclei over two discontinuous Percoll gradients prepared as follows: 5 ml steps containing 60% (v/v) and 35% (v/v) Percoll A: buffer A. Percoll A is made as

follows: 34.23 g sucrose, 1.0 ml 1 M Tris-HCl (pH 7.2), 0.5 ml of 1 M $MgCl_2$, 34 μl of β-mercaptoethanol and Percoll to a final volume of 100 ml.

10. Once the gradients have been loaded, centrifuge at 2000 rpm in a Beckman JS-13 rotor. After 5 min increase speed to 8000 rpm and spin for an additional 15 min. The starch will pellet, the nuclei will band at the 35–65% interface and intact chloroplasts will band at the 0–35% interface. Collect the nuclei from the 35–65% interface and dilute with 5–10 volumes of buffer A.

11. Pellet the nuclei by centrifugation at 8000 rpm in a JS-13 rotor for 10 min. The nuclei can be visualized by light microscopy following staining with 1% Azure in buffer A (without ethidium bromide).

12. Resuspend the nuclei in 5 ml of 250 mM sucrose, 10 mM Tris-HCl (pH 8.0), 5 mM $MgCl_2$ by homogenization.

13. Bring the volume to 20 ml with TE (10 mM Tris-HCl pH 8.0, 1 mM EDTA) and add EDTA (pH 8.0) to a final concentration of 20 mM. Add 1 ml of 20% Sarkosyl (w/v) and Proteinase K to 100 μg/ml. Incubate at 55 °C until the solution clarifies (approximately 2 h).

14. Allow the nuclei preparation to cool to room temperature and add 21 g CsCl. When the CsCl has dissolved, add 1 ml of 10 mg/ml ethidium bromide and mix by gentle inversion. Transfer to two quick-seal tubes and centrifuge in a Beckman Ti 70.1 or equivalent rotor at 65,000 rpm at 20 °C for 16–24 h.

15. Remove the banded DNA with a 15 gauge needle. If the DNA is of high molecular weight the band should be very viscous. Gently extract the ethidium bromide with an equal volume of isopropanol saturated with CsCl. Repeat the extraction until there is no ethidium bromide present in the organic phase.

16. Dialyze against three changes of 1 liter of TE.

17. Concentrate the DNA by ethanol precipitation [1].

Isolation of total DNA [66]

1. Grind 20 g of frozen tissue in 15 ml of DEB buffer (0.2 M Tris-HCl pH 8.0, 100 mM EDTA, 1% (w/v) SDS, 200 μg/ml Proteinase K) and 10 g of acid-washed glass beads (75–150 μm) with a mortar and pestle.

2. Transfer to a 50 ml centrifuge tube and incubate at 48 °C for a minimum of 1 h. The solution should become viscous.

3. Pellet the starch and glass beads by centrifugation at 4,200 rpm in a Beckman JA-14 rotor for 10 min.

4. Transfer the supernatant to a 250 ml centrifuge bottle, add 2 volumes (approximately 60 ml) of ice-cold ethanol and mix by inversion.

5. Pellet the DNA by centrifugation at 8,000 rpm in a Beckman JA-14 or equivalent rotor for 15 min at room temperature.

6. Decant and discard the supernatant and resuspend the pellet in 20 ml of TE (10 mM Tris-HCl pH 8.0, 1 mM EDTA) by gentle shaking on a rotary shaker for approximately 1 h at room temperature. Alternatively, the pellet may be dissolved overnight at 4 °C.

7. Centrifuge the DNA-containing solution at 8,000 rpm in a JA-14 rotor for 15 min at room temperature.

8. Transfer the supernatant to a clean centrifuge bottle and add two volumes of ice-cold ethanol.

9. Pellet the DNA by centrifugation at room temperature for 5 min at 8000 rpm in a JA-14 rotor. Pour off the supernatant and resuspend the pellet in 20 ml of TE.

10. Purify the DNA by CsCl banding as described in the preceding protocol.

Construction of cosmid libraries

The success in constructing a physical map depends largely on the quality of the libraries employed. Therefore, we will discuss in considerable detail the protocols which we use to construct libraries. In particular we describe protocols for making random shear cosmid libraries. The reason for using mechanical shear is to avoid any potential bias which might be introduced by either the non-random distribution of restriction sites or differential kinetics of cleavage when limit restriction digests are used to prepare the inserts. In practice, neither differential cleavage nor the uneven distribution of restriction sites is likely to be the major factor in contributing to library bias. Nonetheless, even a small fraction of the genome which contains regions with a non-random distribution of restriction sites or sites which are differentially cleaved, will create gaps in the map since these sequences will be selectively lost from the population. The advantage of mechanical shear is that shear forces are not expected to respect local sequence variations and should therefore produce a totally random distribution of fragments. On the other hand, the cloning of fragments generated by shearing is unfortunately rather inefficient.

Steps in the procedure

Preparation of inserts
1. Bring 50 to 100 μg of nuclear DNA to a total volume of 500 μl with TE (10 mM Tris-HCl pH 8.0, 1 mM EDTA).
2. Shear the DNA to an average size of 50 to 100 kb. We have found that vortexing the DNA for approximately 1 min at the maximum setting results in a sample with a size average of around 50 to 100 kb (as visualized by ethidium bromide staining following fractionation on a 0.3% agarose gel). The average size can be adjusted by changing both the time and speed of the vortexing step. It may be necessary to optimize the conditions for each DNA preparation. The mean fragment size should be checked by electrophoresis on a 0.3% agarose gel using intact λ DNA as a standard [1].
3. Size-fractionate the sheared DNA on a 36 ml 1.25 to 5.0 M NaCl (w/v NaCl/TE) gradient by centrifugation at 27,000 rpm for 16 h at 18 °C in a Beckman SW27 or equivalent rotor. Alternatively, one can use either a 10–40% sucrose gradient or agarose gel electrophoresis for size fractionation [1, 36]. The sizing step improves the efficiency of the system by minimizing the number of ligation products which are not in the size range for *in vitro* packaging into bacteriophage λ particles. More importantly, size fractionation reduces the potential for generating cosmids harboring sequences which are non-contiguous in the genome.
4. Collect 0.5 ml fractions from the gradient. Check the size distribution by running 15 μl of every third fraction on a 0.3% agarose gel.
5. Pool the fractions having a size distribution between 45 and 70 kb and precipitate with an equal volume of isopropanol.

Note

For all subsequent steps it is important that the samples be handled gently to avoid further shearing of the fragments. Mixing should done by gentle pipetting, never vortexing! In addition, it is often difficult to resuspend large fragments following ethanol precipitation. It may therefore be necessary to allow the pellets to resuspend overnight at 4 °C. Complete drying of the pellets should be avoided since dehydrated pellets are very difficult to resuspend.

6. Dissolve the pellet in 400 μl of TE (10 mM Tris-HCl pH 8.0, 1 mM EDTA), add 200 μl 7.5 M NH$_4$OAc (pH 7.5) and precipitate with 800 μl of ethanol. Wash the pellet with 70% ethanol and briefly air-dry.

T4 polymerase repair of sheared DNA

In order to get efficient ligation of sheared DNA it is necessary to produce blunt ends. There are two steps to this procedure, dephosphorylation with calf intestinal phosphatase (CIP), followed by T4 polymerase 'polishing' of the ends. The dephosphorylation serves two functions: (i) by removing the 5′ phosphates the likelihood of getting unwanted ligation products due to multiple inserts is greatly reduced; and (ii) the removal of the 3′ terminal phosphates is necessary to get efficient polishing of the ends. This is important since 3′ phosphates are inhibitory to T4 polymerase.

1. Bring 5 μg of DNA to 40 μl with TE and add the following:
   - 5 μl of 10× HIN buffer (100 mM Tris-HCl pH 7.5, 600 mM NaCl, 66 mM MgCl$_2$, 10 mM DTT)
   - 5 μl of 1 M Tris-HCl (pH 9.0)
   - 2 μl (20 units) of CIP
   Incubate for 40 min at 37 °C.
2. To terminate the reaction add:
   - 130 μl TE
   - 20 μl 10× STE (100 mM Tris-HCl pH 8.0, 1 M NaCl, 10 mM EDTA)
   - 10 μl 10% SDS
   Incubate at 65 °C for 15 min.
3. Extract 3× with an equal volume of phenol/chloroform (Note: phenol/chloroform is phenol/chloroform/isoamyl alcohol in the ratio 25:24:1). Precipitate with 0.5 volumes of 7.5 M NH$_4$OAc (pH 7.5) and 2 volumes of ethanol. Wash the pellet with 70% ethanol, air-dry for 5 min and dissolve in 40 μl of TE.
4. Add the following:
   - DNA in 40 μl of TE
   - 5 μl of 10× dNTPs (250 μM solution of all four dNTPs)
   - 5 μl of 10× T4 pol buffer (330 mM Tris-OAc pH 7.9, 660 mM KOAc, 100 mM Mg(OAc)$_2$, 5 mM DDT, 10 mg/ml BSA)
   - 1 μl T4 polymerase (2 units)
   Incubate at 37 °C for 30 min.
5. Extract 2× with phenol/chloroform, ethanol-precipitate the aqueous phase, wash the pellet with 70% ethanol and resuspend in 20 μl of TE.

Blunt end ligation

This protocol is based on the observation that the rate of blunt-end ligation can be increased by over three orders of magnitude in the presence of large polymers such as polyethylene glycol (PEG) [46, 68]. Ligations are carried out in the presence of 15% PEG in a total volume of 60 μl. Since PEG-mediated stimulation of the ligation rate occurs over a fairly narrow concentration range [46], a rather large reaction volume is used to minimize errors associated with pipetting viscous PEG solutions. It should be noted that DNA tends to be readily sedimentable in 15% PEG [46] so centrifugation should be avoided.

Vector DNA is prepared by the method described by Ish-Horowicz and Burke [27]. Vector 'arms' are prepared by taking two aliquots of the vector, one of which is cleaved with an enzyme which cuts to the right of the cos site and the other with an enzyme with cleaves to the left of the cos site. The vector arms are then dephosphorylated and cut with an enzyme which generates the blunt-end cloning site. The right and left arms are then purified by agarose gel electrophoresis and eluted from the gel slices by the Gene-Clean procedure (Bio 101). While this method of preparing vector requires more enzymatic steps the efficiency is improved since the dephosphorylation prevents the ligation of tandem vectors and therefore suppresses background due to colonies harboring cosmids with no inserts.

1. To 5 μg of insert DNA in 20 μl of TE add the following:
   - 1 μg of each vector arm
   - 3 μl of 10× ligase buffer (660 mM Tris-HCl pH 7.5 50 mM MgCl$_2$, 50 mM DTT, 10 mM ATP)
   - H$_2$O to 30 μl
2. Add 1 μl of T4 ligase (5 units) and mix by gentle pipetting.
3. Add 30 μl of 30% PEG 8000 in H$_2$O and gently mix.
4. Add 1–2 μl of T4 ligase (5–10 units), mix well by gentle pipetting and incubate at 20 °C for 12 to 24 h.
5. Add 1 μl of 1 μg/μl acrylamide in H$_2$O (carrier) and precipitate with 13 μl of 5 M NH$_4$OAc (pH 7.5) and 200 μl of ethanol.
6. Carefully wash the pellet 2× with 70% ethanol, air-dry for 5 min and resuspend overnight in 10 μl of TE at 4 °C.

In vitro *packaging of cosmids*

Several procedures are available for preparing extracts for the *in vitro* packaging and subsequent introduction of recombinants into host cells [1, 25, 26, 36]. Efficiencies in the range of 10^7-10^8 recombinants/μg can be reproducibly attained. However, as mentioned in the Introduction, many of the strains commonly used for preparing packaging extracts are based on *E. coli* K12 and contain the *hsd* restriction enzyme. In addition, extracts are usually prepared from cells containing *mcr*A and *mcr*B restriction activities, which have the potential to bias the packaging of clones having a high degree of methylation. Bias introduced during packaging can be minimized by preparing extracts from restriction-deficient hosts (*mcr*A$^-$, *mcr*B$^-$, *hsd*R$^-$). Alternatively, extracts are commercially available (Stratagene) which are *mcr*A, *mcr*B, *mrr* and *hsd* restriction-deficient. The commercial extracts also provide high packaging efficiencies (10^9 pfu/μg) and are available in a form which preferentially package recombinants which are 47 to 51 kb in length and therefore maximize the mean insert size.

120

Steps in the procedure
1. Package up to 4 μl of the ligation reaction directly using the appropriate protocol.
2. Store the library in 500 μl of SM (100 mM NaCl, 10 mM MgCl$_2$, 50 mM Tris-HCl pH 7.5, 0.01% (w/v) gelatin) at 4 °C over 20 μl of chloroform.
3. Grow an overnight culture of the bacterial cells in liquid LB medium [1] containing 10 mM MgCl$_2$ and 0.2% (w/v) maltose at 37 °C. We generally use DK1 as the host: (*srl-recA*)δ306, *hsdr*, *ara*D-139, (*ara-leu*)δ7697, (*lac*)δX74, *mcr*A, *mcr*B, *rsp*i-20 [30]. Subculture the overnight culture into LB plus Mg^{2+} and maltose by diluting 1 ml into 50 ml and incubate at 37 °C. Grow to an A$_{600}$ of 1.0, harvest the cells by centrifugation for 5 min at 4,000 g and resuspend the pellet in 10 ml of 10 mM MgCl$_2$.
4. Dilute 5 μl of the library into 100 μl of SM. Add 0.2 ml of the host cells, mix gently and incubate at 37 °C for 20 min. Add 1 ml of LB and incubate at 37 °C for 40 min on a roller drum. Plate varying amounts onto LB plates containing the appropriate antibiotic.

Cosmid DNA miniprep procedure

The miniprep procedure which we describe is based on the alkali lysis method of Birnboim *et al.* [4]. Most of the modifications are intended to simplify the handling of large numbers of samples. This procedure is based on the use of repetitive dispensers and centrifuges which hold racks of microcentrifuge tubes (Eppendorf model 5414 or Beckman model 12). By using labeled tube holders it is unnecessary to label sets of individual tubes and the number of manipulations is minimized since the samples are handled in groups of 10. The use of repetitive dispensers greatly simplifies the addition of reagents. Using this protocol, we typically work up 200 minipreps per day. While this protocol is more time-consuming than procedures where samples are prepared in microtiter plates, it has the advantage that it gives reasonably good yields of relatively pure DNA which can be subsequently used for other purposes such as making probes.

Steps in the procedure
1. Inoculate 3 ml of LB medium, containing the appropriate antibiotic, with a single colony. The colonies should be freshly plated. We never use colonies which have been stored on the plate for more than 5 days.
2. Grow the cultures at 37 °C for 18–22 h on a roller drum.
3. Remove 2.0 ml of the culture into a 2.2 ml Eppendorf tube. Cultures can be poured directly into the tube.
4. Pellet the cells by centrifugation for approximately 1 min in a microcentrifuge at 12,000 g.
5. Remove the supernatant by aspiration with a drawn-out pipette.
6. Resuspend the pellet (vortex for 15 s) in 250 μl of:
   – 50 mM glucose
   – 10 mM EDTA
   – 25 mM Tris-HCl (pH 8.0)
   Incubate on ice for 5 min.

7. Add 250 μl of 0.2 M NaOH, 1% SDS (fresh) and mix by approximately 15 inversions (do not vortex!). Cool on ice for 5 min.

8. Add 200 μl of 3.0 M NaOAc, pH 4.8 (ice-cold) and mix by approximately 15 inversions (do not vortex!)

9. Let sit on ice for 30–60 min.

10. Pellet the debris by centrifugation for 5–15 min.

11. Remove 600 μl of the supernatant into a 1.5 ml Eppendorf tube. Fill the tube with 100% ethanol and mix well.

12. Pellet the DNA by centrifugation for 2–5 min, then decant the ethanol by inverting the rack on a tissue.

13. Briefly air-dry the pellet for 5–10 min then resuspend in 250 μl of TE(5) (Note: the 5 denotes that the TE contains 5 mM EDTA rather than the usual 1 mM. TE(5) is 10 mM Tris-HCl pH 8.0, 5 mM EDTA). Leave at room temperature for 15 min then vortex briefly.

14. Add 250 μl 4.4 M LiCl, mix, and incubate on ice for 30 min.

15. Centrifuge for 5 min to pellet debris, then remove 450 μl of the supernatant to a new tube. Fill tube with ethanol, mix by inversion and place at $-20\,^\circ$C for 20 min.

16. Spin for 2–5 min to pellet the DNA then decant the ethanol.

17. Wash the pellet with 95% ethanol by adding 1 ml of ethanol, centrifuge for 1 to 2 min then decant and discard the supernatant.

18. Briefly air-dry the pellet and resuspend overnight at 4 $^\circ$C in 50 μl of TE. Vortex briefly.

Yield should be 1–3 μg of cosmid DNA.

Note

Although the LiCl precipitation is not essential, it is effective for removing residual protein, cell debris, contaminating *E. coli* DNA and a significant fraction of the RNA. The quality of the DNA is therefore improved, giving a cleaner and more reproducible fingerprint.

Cosmid minipreps in microtiter plates

This protocol is based on the use of 96-well microtiter plates, multi-sample pipetting devices and repetitive dispensers. The batch mode of processing allows large numbers of samples to be handled simultaneously with minimal effort. It is convenient to prepare either 2 or 4 plates simultaneously, 192 and 384 clones respectively, although more can be processed if needed. This protocol is well suited for procedures where small quantities of DNA from a large number of samples are required. When larger quantities of DNA are required the clones can easily be retrieved from the glycerol stocks. The key to this protocol is growing the bacterial cultures to a high density in 96-well microtiter dishes. One problem frequently encountered is that the cultures grow poorly due to poor aeration. We describe two alternatives: growing the cultures in 1 ml Micronic tubes (Flow Laboratories) which are held in boxes in a standard 96-well microtiter dish configuration [19] and growing the cells in NO_3-containing medium (B. Seed, personal communication). The use of 1 ml tubes permits the shaking of the cultures in an orbital incubator, since the dimensions of the tubes reduce the likelihood of cross-contamination while providing

122

adequate aeration. For the NO_3-containing medium, cultures are grown in standard microtiter plates. The nitrate serves as the terminal electron acceptor for anaerobic respiration thereby eliminating the need for aeration.

Growth of cultures in NO_3 medium
Dispense 300 µl of TYGP medium containing the appropriate antibiotic into the wells of a round-bottom 96-well microtiter plate. Transfer the colonies with a sterile toothpick into each well, with the toothpicks being left in the wells until the entire plate has been inoculated. The plates are covered to prevent evaporation and incubated at 37 °C for 16 to 24 h.
– TYGP medium:
  – 20 g Bacto-tryptone
  – 10 g yeast extract
  – 10 ml 80% glycerol
  – 5 g Na_2HPO_4
  – 10 g KNO_3
  – 1 l H_2O

Growth of cultures in Micronic tubes
One ml Micronic tubes are labeled if necessary and autoclaved in their boxes. A 0.5 ml portion of LB medium containing antibiotic is dispensed into each tube. Colonies are transferred with sterile toothpicks. The lid of the box is attached and the boxes are arranged in an orbital incubator at approximately a 30° angle to facilitate aeration and resuspension of the cells. Cultures are grown at 37 °C with shaking at 300 to 500 rpm for 16 to 24 h.

Alkali lysis in microtiter plates
It is convenient to process plates as pairs or in sets of 4 for balancing during centrifugation. The following protocol is based on the use of 12 tip multiple pipettes, although 96 tip pipettes could be used to further speed up the process. The disposable tips are in autoclavable boxes in groups of 96. When sterility is not an issue, it is both cheaper and more convenient to re-use the tips by flushing them with deionized water.

Steps in the procedure
1. For cultures grown in Micronic tubes, transfer 250 µl into a round-bottom microtiter plate. When NO_3 medium is used for growth, the plate is used directly following removal of 50 µl for glycerol stocks.
2. Spin the saturated cultures for 10 min at 2500 rpm in a centrifuge having microtiter plate carriers (Sorval RT6000). Immediately decant the medium by flicking the inverted plate then drain briefly by inverting the plates on tissues.
3. Loosen the pellets by vigorous vortexing.
4. Add the following;
  – 50 µl 10 mM EDTA, 1 mg/ml BSA
  – 100 µl 1% SDS, 0.2 M NaOH (prepared fresh or protected from atmospheric CO_2)

    – 50 μl 2.5 M KOAc, 2.5 M HOAc (250 g/150 ml to 1 l)

5. Cover the plate and agitate vigorously on a rotary shaker.

6. Centrifuge at 2500 rpm for 5 min.

7. Remove 200 μl from each well and transfer to a new microtiter plate. Do not try to recover all of the supernatant. Place the pipette tip at the edge of the wells where the curvature of the bottom begins.

8. Add 150 μl of isopropanol to each well. Cover the plate, mix by agitation on a rotary shaker and place at −20 °C for 30 min to 1 h.

9. Centrifuge the plates at 2500 rpm for 25 min. Decant the supernatant and drain the plates on tissues for 5 min.

10. Resuspend the pellet in 25 μl of TE. Add 25 μl of 4.4 M LiCl, mix and place at 4 °C for a minimum of 1 h.

11. Centrifuge at 2500 rpm for 15 min and transfer 40 μl of the supernatant to a new plate containing 100 μl of ethanol per well.

12. Mix and place at −20 °C for 1 h.

13. Spin at 2500 rpm for 10 min, decant the supernatant by flicking the inverted plate and drain well.

14. Wash the pellet with 200 μl of 95% ethanol, spin, discard ethanol and air-dry the plates for at least 10 min. The dried pellets are resuspended in 10 to 20 μl of TE. Samples are stored at −20 °C with the wells sealed with parafilm.

Fingerprinting

*Fingerprint reactions with Hin*d III/*Sau*3A

In the protocol which we describe, the clones are digested with *Hin*d III and the resultant ends are simultaneously labeled with reverse transcriptase and the appropriate nucleoside triphosphates. Following thermal inactivation, the samples are then cleaved with a second enzyme, *Sau*3A. The protocol may be modified for any enzyme or combination of enzymes. There are several considerations for choosing an enzyme(s): the enzyme(s) should be chosen such that the average number of labeled bands is optimal for the statistical detection of overlaps [32]. When the inserts are prepared by partial digestion with a restriction enzyme it may be desirable to maintain the same cleavage specificity in the fingerprinting reaction to avoid anomalous bands arising from the insert/vector junction. In practice, this is not important when the fingerprint is composed of a large number of bands. The enzymes used should be active in a single buffer to minimize the number of manipulations required. Preferably, restriction enzymes should be used which retain activity during extended incubation and which are readily available at high concentration. The former minimizes problems associated with analyzing gels containing partial digestion products, while the use of concentrated enzymes eliminates potential glycerol effects (i.e., inhibition of activity and star activity) [43, 47]. It may be further advisable to avoid restriction enzymes which are know to cleave their recognition sequences at significantly different rates [20, 42, 65]. Differences in the order of 50-fold have been observed for several enzymes. Differential kinetics of cleavage can contribute to differential labeling and to partial digests, both of which can complicate data analysis. On the other hand, if the differential labeling of sites is reproducible, differences in band intensity can be exploited when assigning overlaps [5].

- Enzyme cocktail for 48 clones:
  - 10 μl ^{32}P-dATP (3000 Ci/mmol)
  - 80 μl water
  - 20 μl 10× HIN buffer (100 mM Tris-HCl pH 7.5, 600 mM NaCl, 66 mM MgCl$_2$, 10 mM DTT)
  - 2 μl RNase (10 mg/ml RNase IA in 10 mM Tris-HCl pH 7.6, 15 mM NaCl, boiled for 15 min)
  - 10 μl 1 mM ddGTP

Steps in the procedure
1. Pre-cool the enzyme cocktail on ice, then add the following:
   2 μl *Hin*d III (50—80 units)
   2 μl M-MLV reverse transcriptase (400 units)
2. Add 2 μl of enzyme cocktail into the wells of a pre-cooled microtiter dish (Nuclon 72 × 10 μl wells) using a Hamilton PB600-1 repetitive dispenser fitted with a disposable tip.
3. Add 0.5 to 1 μl (25—50 ng) of the cosmid mini-prep DNA to each well.
4. Seal the micotiter dish with a glass plate which has been covered with parafilm to insure a tight seal. Incubate at 37 °C for 45 min.

126

5. Heat-kill the reaction for 30 min at 68 °C. Following the heat inactivation, cool the microtiter dish on ice.
   - Sau3A cocktail:
     - 200 μl water
     - 20 μl 10× HIN buffer (100 mM Tris-HCl pH 7.5, 600 mM NaCl, 66 mM MgCl$_2$, 10 mM DTT)
     - 50–100 units of Sau3A. Volume should be less than 8 μl to avoid glycerol effects.
6. Add 4 μl of the Sau3A cocktail to each well using a Hamilton PB600-1 repetitive dispenser.
7. Re-seal the dish and incubate at 37 °C for 2–3 h.
8. Stop the reaction by addition of 5 μl of formamide dye to each well (formamide plus 10 mM ETDA and tracking dyes).
9. To an empty well add 1 μl of the labeled markers (see below) to 10 μl formamide-dye mix.
10. Place the microtiter dish at 90 °C for 8 min. Note: the microtiter dish should be left uncovered.

λ Sau3A markers
1. Add the following:
   - 20 μl water
   - 5 μl 10× HIN buffer
   - 15 μl ^{35}S-dATP (500 Ci/mmol)
   - 6 μl Sau3A-digested λ DNA (0.5 μg/μl)
   - 2 μl 10 mM dGTP
   - 2.5 μl 10 mM ddTTP
   - 1 μl M-MLV reverse transcriptase (200 units)
2. Incubate at 37 °C for 30 min. Add EDTA to 10 mM and store at −20 °C.

Fingerprinting gels
Since the gels are run with ^{35}S-labeled markers, it is necessary to fix and dry the gels prior to autoradiography. We dry the gel directly onto the glass plate. Alternatively, the gels may be fixed, transferred to 3 MM paper and dried on a gel dryer. However, binding the gel directly to the glass plate has the advantage that it prevents distortion of the sample wells. To form the wells we use combs with 60 usable slots which are 4 mm wide and separated by 1 mm. The 1 mm separation between wells is close to the minimal distance which still gives reproducible polymerization. To insure that the wells form properly the combs are de-gassed and then flooded with N$_2$ gas, since the level of oxygen present in the pores of the comb is often sufficient to inhibit polymerization of the narrow slots.

Pre-treatment of gel plates
Siliconize the larger of the two plates with Sigma coat (dichlorodimethylsilane), by spreading the concentrated solution onto the plate. Let the solution air-dry for approximately 5 min, then remove the excess with 70% ethanol. The second plate

is treated with methacryloxypropyltrimethoxysilane, which covalently binds the gel to the glass plate. The binding silane is prepared by adding 5 μl of methacryloxy-propyltrimethoxysilane to 3 ml of ethanol plus 50 μl of 10% acetic acid. The binding silane is spread directly on the glass plate with a tissue, air-dried for 5−10 min and the excess is removed by washing extensively with ethanol. Note: The methacryloxypropyltrimethoxy silane is insoluble in water, but it is soluble in acrylamide. It is therefore very important to remove all of the excess binding silane, since failure to do so can result in binding to both plates.

Steps in the procedure
1. Gels are 4% acrylamide, Tris/borate/EDTA, 8 M urea.
   - 1 gel
   - 48 g urea
   - 10 ml 40% acrylamide (19:1 acrylamide/bisacrylamide)
   - 10 ml 10× TBE (500 mM Tris-borate pH 8.3, 10 mM EDTA)
   - 44 ml H$_2$O
2. Filter the gel mix to remove any insoluble material.
3. To each 100 ml of gel mix add:
   - 200 μl TEMED
   - 200 μl of 10% ammonium (w/v) persulfate
4. Pour gel and allow to polymerize for at least 1 h prior to running.
5. Load 1 μl of sample per well and 0.5 μl *Sau*3A markers every 7th well.
6. Run at 45 mA (approximately 1600 V) until the bromophenol blue dye is approximately 2.5 cm from the bottom of the gel.
7. Fix the gel for 15 min in 1 l of 10% acetic acid and then rinse for 15 min in 2 l of water.
8. The gels are dried directly onto the glass plate in a drying oven for 15 to 30 min at 80 °C. Alternatively, the gels may be dried overnight at room temperature.
9. Autoradiograph for one to several days on Kodak XAR5 film. The exposure time should be determined empirically.
10. The gels are removed from the glass plate by soaking in 20% Countoff (NEN) or a solution of 1% NaOH.

Image analysis of fingerprint autoradiograms
The software which has been developed to assist in mapping by fingerprint analysis has been described in detail [13, 62, 63]. Briefly, input data are attained using a scanning densitometer and an image processing package. The procedure for image processing involves a preliminary densitometric pass to locate band-like features, lane tracking, a precise densitometric pass and alignment of the marker bands with the standard. Following alignment of the markers, a normalized grid is calculated by linear interpolation between nearest markers and used to calculate the band positions for each lane. For interactive editing the band positions are displayed as colored lines superimposed on an image of the autoradiogram. A VAX station II/GP4 (Digital) is used for the display and editing of the data (B. Hauge, W. Loos and H. Goodman, unpublished). The bands are displayed over the marker lanes, together with the

bands from a single sample. Using the 'mouse' the operator can selectively remove unwanted bands before moving to the next sample lane. As individual lanes are edited, the normalized position of the bands are written to a data base. Clone matching and contig assembly are performed as described [13, 62].

Library screening

It is not expected that a complete map can be assembled based solely on random clone analysis. At some point it is necessary to close the gaps by selecting the missing clones. There are several alternatives for selecting the linking clones by hybridization. The choice of strategies depends to a large extent on the particular project and the nature of the genome being analyzed. We describe two approaches: the selection of linking clones with riboprobes from the ends of existing contigs and using YAC clones to probe a representative collection of cosmids.

Construction of YAC libraries

Recent developments allow the routine cloning of 0.5 Mb-sized DNA fragments as artificial chromosomes in yeast [6, 7]. This is an improvement of at least an order of magnitude over previously existing techniques. The construction of YAC libraries involves the ligation of large DNA fragments (50–1000 kb) into vectors containing selectable markers and the functional components of a eukaryotic chromosome [39; see also Albertsen *et al.*, this book]. The constructs are transformed into *S. cerevisiae* where they are replicated with the host chromosomes. Successful construction of a YAC library depends to a large extent on the ability to isolate megabase sized DNA molecules for the preparation of inserts.

The procedures given here have been used for the construction of *Arabidopsis* YAC libraries. A more detailed discussion of the construction and application of YAC libraries is presented in this volume by Albertsen *et al.*

Isolation of Mb-sized DNA from protoplasts

The DNA isolation procedure which we describe is based on the isolation of protoplasts which are subsequently embedded in low-gelling agarose (I. Hwang and H.M. Goodman, unpublished; also see van Daelen and Zabel, this book). The samples are handled in gel plugs to minimize breakage due to shear forces [7, 64]. The gel inserts are treated with a combination of detergents and enzymes which remove cell membranes, RNA and proteins leaving essentially naked DNA. A high concentration of EDTA is used to inactivate cellular nucleases and an extensive proteinase K treatment in the presence of detergents is used to remove proteins.

Steps in the procedure
1. Harvest 50 g of tissue which has been destarched by placing the plants in the dark for 48 h.
2. Wash with ice-cold water and cut into small pieces using either a single-edge razor blade or scissors.
3. Add 500 ml of protoplast buffer (2% cellulase, 0.25% macerozyme, 0.5 M mannitol, 8 mM $CaCl_2$) and place on a rotary shaker. Incubate overnight at room temperature with shaking at 120 rpm.

4. Filter the homogenate through a sieve with 180 μm pores, then through a second sieve with 75 μm pores. Clearly the appropriate pore size is dictated by the nuclear volume and must be adjusted accordingly.

5. Harvest the protoplasts by centrifugation at room temperature for 5 min at 3000 rpm in a JS-13 or equivalent rotor.

6. Resuspend the pellet in 100 ml of 0.5 M mannitol, 8 mM $CaCl_2$. Harvest the protoplasts by centrifugation for 5 min in a JS-13 rotor.

7. Repeat the wash described in step 6.

8. Resuspend the pellet in 10 ml of 0.5 M mannitol and incubate at 37 °C for 5 min. Add 7 ml of 2% low-melting-point agarose prepared in 0.5 M mannitol which is held at 45 °C. Mix thoroughly and allow to solidify.

9. Cut the agarose block into small pieces and incubate overnight at 45 °C with 2.5 μg/ml proteinase K in 0.5 M EDTA, 20 mM Tris-HCl (pH 8.0), 2% sarcosyl.

10. Wash the agarose pieces extensively with 10 mM EDTA, 20 mM Tris-HCl (pH 8.0) at room temperature and store at 4 °C.

Cloning in YAC vectors [6]

1. To establish the conditions for partial digestion of high-molecular-weight DNA set up a series of tubes containing approximately 1 μg of agarose embedded DNA per tube. Add serial dilutions of the restriction enzyme in the appropriate buffer which has been prepared without Mg^{2+} (the Mg^{2+} is required for cleavage).

2. Allow the enzyme to diffuse into the gel slice by incubating at 37 °C for 3 h.

3. Add Mg^{2+} to a final concentration of 6 mM and continue the incubation at 37 °C for 1 h.

4. To terminate the reaction add 0.5 M EDTA (pH 8.0) to a final concentration of 20 mM and incubate at 65 °C for 10 min.

5. The samples are analyzed by CHEF gel electrophoresis using yeast chromosomes and λ ladders as the size standards [11]. Electrophoresis is through a 1% agarose gel in 0.5× TBE at 13 °C. The gel is run for 20 h at 200 V using a 60 s switch interval.

6. Photograph the gel and determine the amount of enzyme needed to produce the maximum fluorescence in the 0.5 to 1 Mb range.

7. Following optimization of the digestion conditions, the reaction is scaled up for 20 μg of DNA in a 200 μl agarose plug.

8. Melt the agarose plug by incubating at 65 °C for 5 min then hold at 37 °C. Add a 100-fold molar excess of the restricted, dephosphorylated pYAC 4 [7] vector.

9. Add 50 μl of 5× ligase buffer (250 mM Tris-HCl pH 7.4, 50 mM $MgCl_2$, 50 mM DTT, 5 mM spermidine, 5 mM ATP, 500 μg/ml BSA) and 20 units of T4 ligase. Mix well and incubate overnight at room temperature.

10. Separate the unligated vector DNA and small molecules by electrophoresis on a 1% low-melting-point agarose gel run for 10 h at 40 V.

11. The large DNA molecules which remain near the origin of the gel are excised

and embedded in a second 1% low-melting gel. The ligation products are then size-fractionated by electrophoresis on a field inversion gel [8]. Electrophoresis is carried out at 200 V for 15 h at 14 °C using a 3 s forward pulse and a 1 s reverse pulse.

12. Slices of the gel containing DNA fragments greater than 100 kb are excised for subsequent transformation.

Yeast transformation [24]

1. Inoculate 10 ml of YEPD medium (1% yeast extract, 2% bacto-peptone, 2% glucose) from a single colony of AB1380, *MAT*α, *ura*3-52, *trp*1, *ade*2-1, *can*1-100, *lys*2-1, *his*5 [7]. Incubate at 30 °C for 18–24 h.
2. Subculture 1 ml of the overnight culture into 80 ml of YEPD medium and grow to a density of 10^7 cells/ml.
3. Harvest the cells by centrifugation at 4,000 *g* for 5 min and wash 2× with 50 ml of 1 M sorbitol.
4. Resuspend the pellet in 10 ml of SCEM (1 M sorbitol, 0.1 M sodium citrate pH 5.8, 10 mM EDTA, 30 mM β-mercaptoethanol) and add 100 μl of 10 mg/ml of zymolyase.
5. Incubate at 30 °C for 15 min with gentle shaking. Test for spheroplasting by adding one drop of the cell suspention to each of 2 tubes containing either 1 ml of 1 M sorbitol or 1% SDS in water. The spheroplasts will lyse in the 1% SDS, while cells containing an intact cell wall will not. Continue incubation until spheroplasting is evident.
6. Melt the agarose block containing the ligated DNA at 65 to 70 °C for 5 min.
7. To 100 μl of spheroplasted cells add 5 μg of carrier DNA and 10 μl of the ligated DNA in the melted agarose. Incubate at room temperature for 10 min.
8. Add 1 ml of 20% (w/v) PEG, 10 mM $CaCl_2$, 10 mM Tris-HCl (pH 7.5). Mix gently and incubate at room temperature for an additional 10 min.
9. Harvest the cells by centrifugation at 3000 *g* for 4 min at room temperature.
10. Resuspend the pellet in 150 μl of SOS medium (1 M sorbitol, 0.25% (w/v) yeast extract, 0.5% (w/v) peptone, 10 μg/ml of uridine and tryptophan, 20 μg/ml of adenine, hisitidine and lysine) and incubate at 30 °C for 20 to 40 min.
11. Add 8 ml of top agar which is held at 48 °C, mix by vortexing and spread onto a pre-warmed agar plate. Pre-warming the plates to 37 °C facilitates uniform spreading of the top agar.
    Top agar and agar plates [24]: 2% agar (w/v), 1.0 M sorbitol, 0.67% (w/v) nitrogen base without amino acids (Difco), 20 mg/ml tryptophan, 10 mg/ml adenine, 20 mg/ml histidine, 20 mg/ml lysine.
12. Incubate the plates at 30 °C for 3 to 5 days.
13. Individual colonies are picked onto agar plates of complete medium lacking uracil and tryptophan which has been supplemented with canavanine [7, 24]. Canavanine resistance selects against *ochre* suppression due to the *sup*-4 gene harbored on the pYAC4 stuffer fragment [7].

Complete medium: 0.67% (w/v) nitrogen base without amino acids (Difco); 1.0 mM adenine, alanine, asparagine, aspartate, cysteine, glutamate, glycine, methionine, proline; 2.0 mM leucine, serine, threonine; 0.75 mM isoleucine, phenylalanine; 0.5 mM tyrosine; 0.2 mM cystine; 0.3 mM histidine; 1.5 mM lysine; 2.5 mM valine. Plates contain 2% agar.

14. The positive clones are then picked into Micronic tubes containing complete medium without uracil and grown to saturation. Glycerol is added to a final concentration of 15% and the clones are held for long-term storage at $-80\,^{\circ}$C.

Small-scale preparation of yeast chromosomal DNA

DNA from recombinant yeast clones is prepared for CHEF gel analysis according to the agarose plug procedure of Burke *et al.* [7, 8].

1. Inoculate cells into 4 ml of complete media [58] lacking uracil and incubate overnight at 30 $^{\circ}$C on a roller drum.
2. Harvest the cells by centrifugation at 4,000 *g* for 5 min.
3. Wash the cells in SCE buffer (1 M sorbitol, 0.1 M sodium citrate pH 7.0, 60 mM EDTA pH 8.0) and resuspend the pellet in 100 μl of SCEM buffer (SCE plus 70 mM β-mercaptoethanol and 2.5 mg/ml zymolyase T20).
4. Heat the cells to 37 $^{\circ}$C for 5 min and add 125 μl of 1.2% low-melting-point agarose in SCE which is held at 42 $^{\circ}$C. Mix by pipetting and pour the mixture into 100 μl polystyrene molds.
5. Incubate the solidified plugs overnight at 37 $^{\circ}$C in a 24-well microtiter plate containing 2 ml of SCEM buffer.
6. Remove the SCEM and replace with 2 ml of lysis solution (0.45 M EDTA pH 8.0, 10 mM Tris-HCl pH 8.0, 1% sarkosyl, 1 mg/ml proteinase K). Incubate at 37 $^{\circ}$C for 12 to 24 h.
7. To determine the insert size and for subsequent isolation, the YACs are separated from the yeast chromosomes by CHEF gel electrophoresis.
8. The plugs can be stored for several months in 500 mM EDTA at 4 $^{\circ}$C.

Library plating

The protocols which we describe apply to both randomly spread colonies and ordered grids. Random clones are spread at a density of about 5000 clones per 15 cm plate. Up to a 1000 clones may be gridded onto a 10 by 8 cm rectangle [14]. Grids can be prepared by either tooth-picking the clones or stamped in a 96-well microtiter configuration using a 96-prong replicator.

Steps in the procedure
1. Spread the required number of colonies, or gridded clones, onto Biotrans nylon membranes which are placed in contact with the appropriate medium: LB plates supplemented with antibiotic for bacterial colonies and complete plates lacking uracil for yeast colonies.

2. Grow colonies overnight; 37 °C for bacterial colonies and 30 °C for yeast colonies.
3. Duplicate filters are prepared as described by Coulson et al. [14].
4. Bacterial clones are disrupted and denatured by stacking the filters between sheets of 3MM paper and autoclaving for 3 min on the fast exhaust cycle. No additional treatment is necessary. We find that this treatment is more effective, as well as much easier, than conventional alkali lysis.

Nylon filters containing yeast colonies are prepared for hybridization as described by Brownstein et al. [6]. Cells are converted to spheroplasts and subsequently lysed by sequentially placing the filters onto the following series of reagent-saturated 3MM paper: lyticase solution (2 mg/ml zymolyase, 1.0 M sorbitol, 0.1 M Na citrate pH 5.8, 10 mM EDTA, 30 mM β-mercaptoethanol) overnight at 30 °C, 10% SDS for 5 min at room temperature, 0.5 M NaOH for 10 min at room temperature and 2× SSC, 0.2 M Tris-HCl (pH 7.5) twice at room temperature. The filters are air-dried for 2 h and irradiated with 1.2 mJ of 260 nm UV light [12].

Riboprobes

This procedure is based on the use of cosmid vectors containing either T3, T7 or Sp6 bacterophage promoters flanking the cloned genomic DNA. Riboprobes are prepared from the ends of existing contigs and used to isolate linking clones. When a large number of joins must be established the RNA probes are prepared from pools of cosmids. By using mixed probes the number of hybridizations is reduced by N, where N is the number of clones used for generating the probes. The pooled clones are most conveniently prepared from the rows of the library matrix. Briefly, the clones from the ends of the contigs and the unattached clones are picked in microtiter dishes and gridded onto nylon filters using a 96-prong replicator. Probes are systematically prepared from rows of clones and hybridized to the ordered grids. Overlaps can be established based on the hybridization data [17] (Fig. 5). The mixed RNA probes are also used to probe different libraries and therefore select clones which are under-represented in the original library.

Steps in the procedure
1. The archived clones are recovered from the glycerol stocks and used to grow overnight cultures in LB containing the appropriate antibiotic.
2. The individual cultures are pooled and used to prepare DNA using the cosmid mini-prep procedure.
3. RNA probes are prepared according to the manufacturer's conditions using T3, T7 or Sp6 (Stratagene) polymerase and ^{32}P-UTP. The reactions are terminated by phenol extraction.
4. The filters are hybridized at 65 °C in 7% SDS, 1 mM EDTA and 250 mM sodium phosphate (pH 7.2) for 12 to 24 h. Pre-hybridization is for 5 min in the same buffer minus the labeled probe [12].

134

5. Washing and autoradiography is as described below except the wash temperature is 65 to 70 °C.

Preparation of cosmid probes by random priming

We routinely prepare labeled probes from cosmids, plasmids, λ clones or purified
fragments using the oligo-labeling procedure of Feinberg and Vogelstein [18]. As
little as 10 ng of DNA is sufficient for labeling, so the procedure is well suited for
making probes from mini-prep DNA.
1. Linearize approximately 50 to 100 ng of cosmid DNA by digestion with the
appropriate restriction enzyme in a total reaction volume in 32 μl.
2. Denature the digested sample by boiling for 5 min.
3. Quick-chill on ice.
4. Add the following:
   – 2 μl of 10 mg/ml BSA
   – 10 μl OLB (see below)
   – 5 μl ^{32}P-dATP (3000 Ci/mmol)
   – 2 units of Klenow fragment
5. Incubate at room temperature for a minimum of 2.5 h. The reactions may be left
overnight.
6. Separate the unincorporated dNTPs on a Sephadex G-50 spin column [1].
7. Prior to hybridization, denature the probe by boiling for 2 min then quick-chill on
ice.
   – OLB is made by mixing solutions A : B : C in the ratio 100 : 250 : 150
     – 'A' 1 ml 1.25 M Tris-HCl (pH 8.0), 125 mM $MgCl_2$, 5 μl of 100 mM dCTP,
dGTP, dTTP
     – 'B' 2 M Hepes pH 6.6 (store at 4 °C)
     – 'C' random hexadeoxyribonucleotides at a concentration of 90
A_{260} units/ml.

Labeling of probes in microtiter plates

The protocol given is for probing 96 filters with YAC clones which are labeled by
random priming [14; Y. Kohara, unpublished]. This protocol can easily be adapted
for samples of isolated DNA such as cosmids. The labeling reactions are done in
96-well microtiter plates and multiple transfers are done with a 12-channel pipette.
The labeled clones are used for cross-probing between the cosmid clones and the
YACs.

Isolation and labeling of YAC clones

1. Separate the YACs from the resident yeast chromosomes by CHEF gel electrophoresis using 1% low-gelling agarose.

2. Cut the YAC clones out of the gel and store at 4 °C until needed.
3. Melt the YAC slices for 5 to 10 min at 70 °C.
4. Add 10 μl of the melted YAC slice to 20 μl of distilled water in Micronic tubes (Flow Labs).
5. Heat to 100 °C for 5 min in a shallow water bath and allow to cool to room temperature. The Micronic tube rack should be covered with aluminum foil during this step.
6. Remove 8 μl into a 96-well microtiter plate containing 4 μl of labeling cocktail. Multiple transfers are performed using a 12-channel pipette.
   - Labeling cocktail for 96 clones:
     - 300 μl OLB (see preceding section)
     - 60 μl 10 mg/ml BSA
     - 60 μl H_2O
     - 25 μl ^{32}P-dATP (3000 Ci/mmol)
     - 150 units of Klenow fragment
7. Seal the microtiter plate and incubate at 37 °C for several hours then incubate overnight at room temperature.
8. Incubate at 70 °C for 5 min in a water bath.
9. To each well add 90 μl of denaturing solution and mix thoroughly by pipetting. Incubate at room temperature for 10 min.
   - Denaturing solution:
     - 3.6 ml 100 mM EDTA (pH 8.0)
     - 1.8 ml 10 mg/ml of denatured salmon sperm DNA
     - 0.9 ml 4 M NaOH
     - 4.5 ml deionized H_2O

Hybridization of filters

- Hybridization solution:
  - 125 mM sodium phosphate (pH 7.2)
  - 250 mM NaCl
  - 10% (w/v) PEG 6000
  - 7% SDS
  - 1% BSA

1. Pipette the labeling reactions into tubes containing 11 ml of the hybridization solution. Using the correct tubes and the appropriate test tube rack, the transfers can be done using a 12 channel pipette.
2. Mix well by inversion and spread the hybridization solution in the lid of a microtiter plate.
3. Soak the filter (DNA side up) in the solution and then invert. If desired add a second filter.
4. Cover the filters with a polythene sheet which has been cut to fit just inside the lid.

5. Stack the lids in an air-tight box and incubate overnight at 68 °C without shaking. The lids are stacked by placing each alternate lid at an angle.

Washing filters

Washing can be done in stainless steel wire baskets which are slightly larger than the filters. By doing so the numeric order of the filters is maintained. Washing is carried out in relatively large volumes with gentle agitation.

1. Wash 2× with 20 mM sodium phosphate (pH 7.2), 5% SDS, 1 mM EDTA for approximately 5 min per wash. The buffer is pre-heated to 68 °C and washing is done on a rotary shaker at room temperature.
2. Wash 6× in 20 mM sodium phosphate (pH 7.2), 1% SDS, 1 mM EDTA for 5 min per wash. The wash buffer is pre-heated to 50 °C.
3. Wash 1× in 3 mM Tris-base at room temperature.
4. Order the filters on sheets of damp 3MM paper and cover with saran wrap.
5. Autoradiograph at −80 °C with an intensifying screen.
6. The filters can be stripped for re-probing by incubating in 2 mM Tris-HCl (pH 8.3), 2 mM EDTA, 0.2% SDS at 70 °C for 10 min with gentle agitation. The filters are stored at 4 °C in the same buffer. If the filters are stored for long periods of time we recommend that the storage buffer be replaced with fresh buffer ever couple of months. Using this treatment it is possible to re-use the filters for a minimum of 20 probings.

References

1. Ausubel FM, Brent R, Kingston RE, Moore DD, Seidman JG, Smith JA, Struhl K, eds (1987) Current Protocols in Molecular Biology. New York: Wiley.
2. Bassett CL, Kushner J (1984) Exonucleases I, III, and V are required for stablity of ColE1-related plasmids in *Escherichia coli*. J Bacteriol 157: 661–664.
3. Bellet AJD, Busse HG, Baldmin RL (1971) in: Hershey AD (ed), The Bacteriophage Lambda, pp. 501–557. Cold Spring Harbor, NY: Cold Spring Harbor Laboratory Press, New York, 501–557.
4. Birnboim HC, Doly J (1979) A rapid alkaline extraction procedure for screening recombinant plasmid DNA. Nucleic Acids Res 7: 1513–1523.
5. Brenner S, Livik KJ (1989) DNA fingerprinting by sampled sequencing. Proc Natl Acad Sci USA 86: 8902–8906.
6. Brownstein BH, Silverman GA, Little RD, Burkem DT, Korsmeyer SJ, Schlessinger D, Olson MV (1989) Isolation of single-copy human genes from a library of yeast artifical chromosome clones. Science 244: 1348–1351.
7. Burke DT, Carle GF, Olson MV (1987) Cloning of large segments of exogenous DNA into yeast by means of artificial chromosome vectors. Science 236: 806–812.
8. Carle GF, Olson MV (1985) An electrophoretic karyotype for yeast. Proc Natl Acad Sci USA 82: 3756–3760.
9. Carrano AV, Lamerdin J, Ashworth LK, Watkins B, Branscomb E, Slezak T, Raff M, de Jong PJ, Keith D, McBride L, Meister S, Kronick M (1989) A high-resolution fluorescence-based, semiautomated method for DNA fingerprinting. Genomics 4: 129–136.
10. Chang C, Bowman JL, DeJohn AW, Lander ES, Meyerowitz EM (1988) Restriction fragment length polymorphism linkage map for *Arabidopsis thaliana.* Proc Natl Acad Sci USA 85: 6856–6860.
11. Chu G, Vollrath D, Davis RW (1986) Seperation of large DNA molecules by contour-clamped homogeneous electric fields. Science 324: 1582–1585.
12. Church GM, Gilbert W (1984) Genomic sequencing. Proc Natl Acad Sci USA 81: 1991–1995.
13. Coulson A, Sulston J, Brenner S, Karn J (1986) Toward a physical map of the nematode *Caenorhabditis elegans*. Proc Natl Acad Sci USA 83: 7821–7825.
14. Coulson A, Waterston R, Kiff J, Sulston J, Kohara Y (1988) Genome linking with yeast artificial chromosomes. Nature 335: 184–186.
15. Cross SH, Little PFR (1986) A cosmid vector for systematic chromosome walking. Gene 49: 9–22.
16. Estelle MA, Somerville CR (1986) The mutants of *Arabidopsis*. Trends Genet 2: 89–93.
17. Evans GA, Lewis KA (1989) Physical mapping of complex genomes by cosmid multiplex analysis. Proc Natl Acad Sci USA 86: 5030–5034.
18. Feinberg AP, Vogelstein B (1983) A technique for radiolabeling DNA restriction endonuclease fragments to high specific activity. Anal Biochem 132: 6–13.
19. Gibson TJ, Sulston JE (1987) Preparation of large numbers of plamid DNA samples by the alkaline lysis method. Gene Anal Technol 4: 41–44.
20. Gingeras TR, Brooks JE (1983) Cloned restriction/modification system from *Pseudomonus aeruginosa*. Proc Natl Acad Sci USA 80: 402–406.
21. Gruenbaum Y, Naveh-Many T, Cedar H, Razin A (1981) Sequence specificity of methylation in higher plant DNA. Nature 292: 860–862.
22. Hamilton RH, Kunsch U, Temperli A (1972) Simple rapid procedures for isolation of tobacco leaf nuclei. Anal Biochem 49: 48–57.
23. Hauge BM, Giraudat J, Hanley S, Hwang I, Kochi T, Goodman HM (1990) Physical mapping of the *Arabidopsis* genome and its application. In: Herman RG (ed), Plant Molecular Biology. New York: Plenum (In press).
24. Hinnen A, Hicks JB, Fink GR (1978) Transformation of yeast. Proc Natl Acad Sci USA 75: 1929–1933.

138

25. Hohn B (1975) DNA as a substrate for packaging into bacterophage lambda, *in vitro*. J Mol Biol 98: 93–106.
26. Hohn B (1979) *in vitro* packaging of λ and cosmid DNA. Meth Enzymol 68: 299–309.
27. Ish-Horwicz D, Burke JF (1981) Rapid and efficient cosmid cloning. Nucleic Acids Res 9: 2989–2998.
28. Kohara Y, Akiyama K, Isono K (1987) The physical map of the whole *E. coli* chromosome: Application of a new strategy for rapid analysis and sorting of a large genomic library. Cell 50: 495–508.
29. Koornneef M (1987) Linkage map of *Arabidopsis thaliana* (2n = 10). In: O'Brien SJ (ed), Genetic Maps, pp. 742–745. Cold Spring Harbor, NY: Cold Spring Harbor Laboratory Press.
30. Kurnit DM (1989) *Escherichia coli recA* deletion strains that are highly competent for transformation and for *in vivo* phage packaging. Gene 82: 313–315.
31. Kuspa A, Vollrath D, Cheng Y, Kaiser D (1989) Physical mapping of the *Myxococcus xanthus* genome by random cloning in yeast artificial chromosomes. Proc Natl Acad Sci USA 86: 8917–8921.
32. Lander ES, Waterman MS (1988) Genomic mapping by fingerprinting random clones: a mathematical analysis. Genomics 2: 231–239.
33. Leach DRF, Stahl FW (1983) Viability of λ phages carrying a perfect palindrome in the absence of recombination nucleases. Nature 305: 448–450.
34. Leutwiler LS, Hough-Evans BR, Meyerowitz EM (1984) The DNA of *Arabidopsis thaliana*. Mol Gen Genet 194: 15–23.
35. Lilley DMJ (1981) *In vivo* consequences of plasmid topology. Nature 292: 380–382.
36. Maniatis T Fritsch EF, Sambrook J (1982) Molecular Cloning: A Laboratory Manual. Cold Spring Harbor, NY: Cold Spring Harbor Laboratory Press.
37. Meselson M, Yaun R (1968) DNA restriction enzyme from *E. coli*. Nature 217: 1110–1114.
38. Meyerowitz EM (1987) *Arabidopsis thaliana*. Annu Rev Genet 21: 93–111.
39. Murry AW, Szostak JW (1983) Construction of artifical chromosomes in yeast. Nature 305: 189–193.
40. Nam H-G, Giraudat J, den Boer B, Moonan F, Loos WDB, Hauge BM, Goodman HM (1989) Restriction fragment length polymorphism linkage map of *Arabidopsis thaliana*. Plant Cell 1: 699–705.
41. Nater WF, Isenberg G, Sauer HW (1986) Structure of *Physarum* actin gene locus *ard*A: a nonpalindromic sequence causes inviability of phage lambda and *rec*A-independent deletions. Gene 48: 133–144.
42. Nath K, Azzolina BA (1981) in: Chirikjian JG (ed), Gene Amplification and Analysis, Vol 1, pp. 113–128. NY: Elsevier-North Holland, New York.
43. Nasri M, Thomas D (1986) Relaxation of recognition sequence of a specific endonuclease *Hind* III. Nucleic Acids Res 14: 811–821.
44. Ochman H, Gerber AS, Hartl DL (1988) Genetic applications of an inverse polymerase chain reaction. Genetics 120: 621–623.
45. Olson MV, Dutchik JE, Graham MY, Brodeur GM, Helms C, Frank M, MacCollin M, Scheinman R, Frank T (1986) Random-clone strategy for genomic restriction mapping in yeast. Proc Natl Acad Sci USA 83: 7826–7830.
46. Pheiffer BH, Zimmerman SB (1983) Polymer-stimulated ligation: enhanced blunt- or cohesive-end ligation of DNA or deoxyribooligonucleotides by T4 DNA ligase in polymer solutions. Nucleic Acids Res 11: 7853–7871.
47. Polisky B, Greene P, Garfin DE, McCarthy BJ, Goodman HM, Boyer HW (1975) Specificity of substrate recognition by the *Eco*RI restriction endonuclease. Proc Natl Acad Sci USA 72: 3310–3314.
48. Pruitt RE, Meyerowitz EM (1986) Characterization of the genome of *Arabidopsis thaliana*. J Mol Biol 187: 169–183.
50. Raleigh EA, Wilson G (1986) *Escherichia coli* K-12 restricts DNA containing 5-methylcytosine. Proc Natl Acad Sci USA 83: 9070–9074.
51. Raleigh EA, Murry NE, Revel H, Blumenthal RM, Westaway D, Reith AD, Rigby PWJ, Ellai

J, Hanahan D (1988) *Mcr*A and *Mcr*B restriction of some *E. coli* strains and implications for gene cloning. Nucleic Acids Res 16: 1563–1575.

52. Raleigh EA (1987) Restriction and modification *in vivo* by *E. coli* K12. Meth. Enzymol 152: 130–141.
53. Roberts RJ (1987) Restriction enzymes and their isoschizomers. Nucl Acids Res 15: r189–r217.
54. Rosenberg SM (1985) *Eco*K restriction during *in vitro* packaging of coliphage lambda DNA. Gene 39: 313–315.
55. Rosenberg SM, Stahl MM, Kobayashi I, Stahl FW (1985) Improved *in vitro* packaging of coliphage lambda DNA: a one-strain system free from endogenous phage. Gene 38: 165–175.
56. Rosenberg SM (1987) Improved *in vitro* packaging of λ DNA. Meth Enzymol 153: 95–102.
57. Shapiro HS (1976) Handbook of Biochemistry and Molecular Biology. CRC Press, pp. 258–262.
58. Sherman F, Fink GR, Hicks JB (1983) Methods in yeast genetics. Cold Spring Harbor, NY: Cold Spring Harbor Laboratory.
59. Signer ER, Weil J (1968) Recombination in bacteriophage λ, I. Mutants defective in general recombination. J Mol Biol 34: 261–271.
60. Smith CL, Econome JG, Schutt A, Klco S, Cantor CR (1987) A physical map of the *Escherichia coli* K12 genome. Science 236: 1448–1453.
61. Smith LM, Sanders JZ, Kaizer RJ, Hughes P, Dodd C, Connell CR, Heiner C, Kent SBH, Hood LE (1986) Fluorescence detection in automated DNA sequence analysis. Nature 321: 674–679.
62. Sulston J, Mallett F, Staden R, Durbin R, Horsnell T, Coulson A (1988) Software for genome mapping by fingerprinting techniques. Comput Applic Biosci 4: 125–132.
63. Sulston J, Mallett F, Durbin R, Horsnell T (1989) Image analysis of restriction enzyme fingerprint autoradiograms. Cabios 5: 101–106.
64. Schwartz DC, Cantor C (1984) Separation of yeast chromosome-sized DNAs by pulsed field gradient electrophoresis. Cell 37: 67–75.
65. Thomas M, David RW (1975) Studies on the cleavage of bacteriophage lambda DNA with *Eco*RI restriction endonuclease. J Mol Biol 91: 315–328.
66. Watson JC, Thompson WF (1986) Purification and restriction endonuclease analysis of plant nuclear DNA. Methods Enzymol 118: 57–75.
67. Wyman AR, Wolfe LB, Botstein D (1985) Propagation of some human DNA sequences in bacterophage λ vectors requires mutant *Escherichia coli* hosts. Proc Natl Acad Sci USA 82: 2880–2884.
68. Zimmerman SB, Pheiffer BH (1983) Macromolecular crowding allows blunt-end ligation by DNA ligases from rat liver or *Escherichia coli*. Proc Natl Acad Sci USA 80: 5852–5856.

7. Construction and application of a YAC library

HANS M. ALBERTSEN, HADI ABDERRAHIM, DANIEL COHEN &
DENIS LE PASLIER
Centre d'Etude du Polymorphisme Humain, 27, rue Juliette Dodu, 75010 Paris, France

Introduction

Genome mapping

As the desire to map genomes has grown over the last decade, the technical means have grown along with it. With the development of different cloning vectors, pulsed-field gel electrophoresis (PFGE) [15, 16, 19, 46 and van Daelen and Zabel, this book], jumping libraries [21, 39] and the polymerase chain reaction (PCR) [43], projects on genome mapping are possible for almost all species. The mapping strategies may be broadly grouped into two general classes, genetic and physical mapping. These two approaches yield qualitatively different information, e.g., genetic maps allow one to get an overall view of an entire chromosome with low resolution of details, whereas physical maps provide very high resolution in local regions of a chromosome without necessarily linking distant regions. The two mapping approaches therefore complement each other, and genome mapping projects including both strategies have good possibilities of being successful.

Physical mapping

Complete physical mapping and eventually sequencing of the human genome was recently initiated with the 'Human Genome Project'. In general, however, research projects are confined to a limited region of a genome. Mapping of a defined region usually starts with genetic mapping of the locus of the trait of interest [6, 8]. Once such a region has been identified, preferably by closely linked polymorphic DNA markers flanking the locus of interest, the goal becomes to physically link the markers. If the distance between two markers has been estimated at 1–2 cM (approximately 1–2 million base pairs in the human physical map), one possible approach for bridging the markers is chromosome walking with yeast artificial chromosomes (YACs) [22, 48, 50]. By using YACs one takes advantage of the fact that very large DNA fragments can be cloned in yeast (*Saccharomyces cerevisiae*). The average insert size in our library is above 400 kb [2], but fragments larger than 1 Mb have successfully been cloned and stably maintained as YACs.

J.S. Beckmann and T.C. Osborn (eds) Plant Genomes: Methods for Genetic and Physical Mapping, 141–165.
© 1992 *Kluwer Academic Publishers.*

The chromosome walk is essentially the same as with other cloned genomic libraries (phage, cosmid), and is initiated by using the genetic DNA markers to identify YACs, in which they are contained. The extremities of each identified YAC are isolated, and these are used as probes to isolate a set of overlapping YACs from the library. The walk is continued in this fashion until the region between the two genetic markers has been bridged with a YAC contig. It is obvious that chromosome walking will be facilitated with increasing YAC size. As YACs are not limited in size in the same way as phage type vectors (e.g. phage λ cosmids and phage P1), but rather by a transformation efficiency inversely related to the insert size, it is likely that YAC libraries with average insert sizes of 800 kb or more may be constructed.

Yeast artificial chromosomes

The YAC vector system, suitable for cloning of very large DNA fragments and hence particularly appropriate in physical mapping and walking experiments, was described in 1987 by Burke *et al.* [14, see also 34, 35]. These authors showed that very large DNA fragments foreign to *S. cerevisiae* can be stably maintained in this host when cloned into a particular vector that mimics a natural yeast chromosome. The vector carries sequences necessary for an artificial chromosome, the centromere (*CEN4*) and an autonomous replicating sequence (*ARS*), which had previously been cloned from *S. cerevisiae* and identified for cell function. The telomeres (*TEL*) were derived from the simple ciliate eukaryote, *Tetrahymena*. As telomeric sequences are homologous between the two species, the yeast telomerase recognizes this template. In addition, the YAC vector carries also selectable markers on each vector arm as well as at the cloning site (see Fig. 1).

The YAC vectors are propagated as bacterial plasmids and, after modification and ligation to the insert DNA, transformed into an appropriate yeast host as a linear molecule. The strain most commonly used as the YAC host is AB1380 (*MATa ade2-1 ura3 can1-100 lys2-1 trp1 his5* ψ^+). This strain transforms with very good efficiency, and allows one to take advantage of all selective features built into the original YAC vectors.

YAC libraries

A series of libraries of various plants and animals have been constructed using YAC vectors [2, 3, 4, 14, 25, 27, 29, 51] with adaptations and modifications particular to each laboratory. The fundamental issues in these papers are the protocols used for extraction of high-molecular-weight insert DNA and how size fractionation was performed. The protocol that we have developed and successfully applied to the generation of a complete human YAC library [2] and that will be described in this chapter includes purification of genomic DNA in

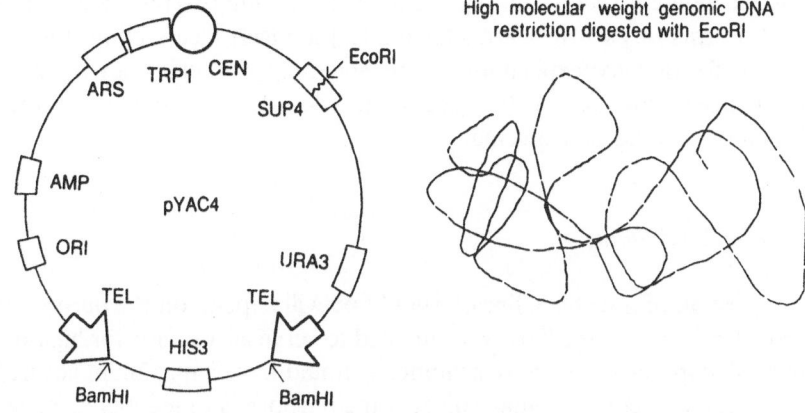

High molecular weight genomic DNA
restriction digested with EcoRI

Functional artificial chromosome

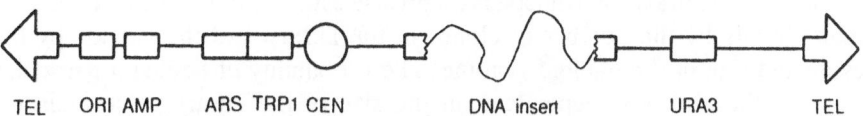

TEL   ORI AMP   ARS TRP1 CEN   DNA insert   URA3   TEL

Fig. 1. Schematic presentation of the pYAC4 vector.

agarose plugs and size fractionation on CHEF pulsed-field gels. A similar
procedure is described by Anand *et al.* [3] and by Hauge and Goodman (this
book). In particular cases specialized libraries have been constructed for isolation of telomeres [5, 10, 17, 23, 41], as well as chromosome-specific libraries
from hybrid cell lines [24, 33, 50].

Construction of a YAC library

The success in YAC cloning is dependent on a number of factors. First of all,
it is essential to extract very high-molecular-weight DNA. The mean size of the
purified fragments should be considerably larger than the fragment size one
wishes to clone. The clonable fragment frequency (F) is a function of the size
of the extracted DNA (D_e) and the desired YAC insert size (D_i). F is the part
of restriction fragments suitable for ligation.

$$F = 1 - \frac{D_i}{D_e}$$

The construction of a functional YAC is further dependent on the attachment
of the two different vector arms (called the left and right arms) to each insert,
because only YACs with a single centromere and two telomeres are functional.
This type of ligation reaction will produce three types of ligation products: one

with two different vector arms ligated to an insert, and two products with either two left arms or two right arms ligated to the insert (see section 'Ligation in agarose' for detailed explications). The actual ligation efficiency is thus 50% of F. If size fractionation is included in the YAC cloning protocol a further loss of material should be expected.

Size of a YAC library

The question of how big a library should be will depend on the purpose of the particular library. If the library is intended to serve as a source for high-density physical mapping of an entire genome, it should have a maximum coverage. A lesser degree of coverage may suffice if the purpose is to identify a single gene or to physically map a relatively small region of the genome. The size of a library is most often expressed in haploid genome equivalents. In order to calculate the size of the library, one must have a precise estimate of the mean YAC size, multiply this by the number of clones in the library and then divide by the estimated size of the haploid genome. The probability of finding a particular clone in the library is dependent on the size of the library. A formula for calculating the probability is shown below [20]:

$$N = \frac{\ln(1-P)}{\ln(1-f)}$$

where N is the number of clones in the library, f is the fractional portion of the genome in a single average-size insert and P is the probability of finding a particular clone in the library.

Plant genomes

The genomic composition of a plant cell falls in three parts: the nuclear genome (one haploid copy containing from 70 million bp to 30 billion bp depending on species), the mitochondrial genome (of the order of a thousand copies ranging from 0.2 million bp to 2 million bp) and the chloroplast genome (approximately one thousand copies each of about 0.15 million bp). The vast amount of DNA in certain plants, much of which is non-coding, repeated sequences, and the polyploid nature of plant genomes further complicates the construction of complete plant YAC libraries thus making physical mapping of plant genomes difficult.

YAC cloning procedures

Extraction of high-molecular-weight DNA

For cloning of large DNA fragments in YACs, it is important that the source DNA is as intact as possible. Extraction of DNA from cells embedded in agarose yields very high-molecular-weight DNA, and therefore is recommended for obtaining DNA suitable for YAC cloning. Still, it must be emphasized that the integrity of the source DNA is crucial for success in YAC cloning; therefore each preparation of DNA should be tested for intregity on PFGE. For extraction of near intact DNA from protoplasts see the protocols by van Daelen and Zabel (this book) and by Cheung and Gale [18].

Protocol for extraction of yeast DNA

Extraction of high-molecular-weight yeast DNA is required for the detailed analysis of YAC clones. Several protocols for extraction of yeast chromosomal DNA have been published [4, 7, 45, 46]. The protocols presented here are in part derived from these protocols.

Dodecyl lithium sulphate (DLS) procedure
This method is used for the purification of intact yeast chromosomes in agarose plugs. The DNA is suitable for both PFGE and restriction enzyme digestion.
1. Cells from a liquid culture are harvested in a 50 ml tube at 900 g and the pellet is rinsed 2 times in distilled water.
2. The yeast cells are resuspended at the desired concentration in 0.5% SeaPlaque agarose dissolved in distilled water and poured into a plug mould on ice.
3. The plugs are transferred to 10 volumes of SCEM with 1 u/ml of Zymolyase 20-T, and incubated at 37 °C for 3 h.
4. The SCEM is replaced with an equal volume of DLS (1% (w/v) dodecyl lithium sulphate, 50 mM NaCl, 10 mM Tris-HCl pH 7.8) and incubated at 50 °C for 2 h whereafter the DLS is replaced with a fresh volume of DLS and the incubation at 50 °C continued for another 2 h.
5. The plugs are rinsed several times in $TE_{10.5}$ (10 mM Tris-HCl pH 7.8, 5 mM EDTA) and stored at 4 °C.

Alternative method: using proteinase K
1. Harvesting and spheroplasting are performed as above (steps 1–3).
2. Replace SCEM with 10 volumes of PKB (0.1 M NaCl, 50 mM EDTA pH 8.0, 0.1 M Tris-HCl pH 8.0, 1% N-lauroylsarcosine, 30 mM β-mercaptoethanol, 0.5 mg/ml proteinase K) and incubate overnight at 50 °C.
3. Rinse several times with $TE_{10.5}$ followed by two rinses of 30 min in $TE_{10.5}$ with freshly prepared PMSF (40 mg/ml in isopropanol) added to 40 μg/ml. After two more rinses of 30 min in $TE_{10.5}$ the plugs are stored at 4 °C.

Extraction of yeast DNA for PCR screening
1. Yeast cultures are grown in 1 ml of the appropriate selective medium in Micronics microtitre dishes at 30 °C with gentle agitation for 3–4 days.
2. The cultures are pooled and the cells are harvested in 50 ml tubes in a centrifuge at 900 *g*. The pellets are resuspended in distilled water and pooled in a single 50 ml tube and harvested.
3. The yeast cells are resuspended at the desired concentration in 0.5% SeaPlaque agarose dissolved in distilled water and poured into a plug mould on ice. For PCR screening of 384 pooled yeast cultures it is convenient to prepare plugs of 3.5×10^9 cells/ml.
4. From the mould the plugs are transferred to 10 volumes of SCEM with 1 unit of Zymolyase 20-T per ml and incubated at 37 °C for 3 h.
5. The SCEM is replaced with an equal volume of DLS and incubated at 50 °C for 2 h whereafter the DLS is replaced with a fresh volume of DLS and the incubation at 50 °C continued for another 2 h.
6. The plugs are rinsed several times in $TE_{10.1}$ (10 mM Tris-HCl, 1 mM EDTA, pH 7.8) and stored at 4 °C.
7. One plug is diluted 10 times with distilled water (resulting in a final DNA concentration of 2.5 ng/μl), boiled for a few minutes and centrifuged in an Eppendorf centrifuge. The supernatant is transferred to a new tube ready for PCR screening.

Restriction digestion of DNA in agarose

DNA purified in agarose is almost intact. This is important for preparation of restriction fragments using rare cutting enzymes or to generate large-sized partial restriction digestion fragments for YAC cloning.

The haploid yeast genome contains approximately 14 million bp. Thus 6.5×10^7 haploid yeast cells contain a total of about 1 μg of DNA. Agarose plugs containing 1–2 μg of yeast chromosomal DNA are convenient for pulsed-field gel electrophoresis (PFGE) of intact yeast chromosomes, as well as traditional gel electrophoresis of restriction-digested yeast DNA.

Partial restriction digestion

YAC libraries constructed from partially restriction-digested DNA are convenient for isolation of overlapping YAC clones and thus for establishing contigs. Several techniques have been developed to generate partially restriction-digested DNA. These include digesting the DNA in the presence of limited amounts of enzyme or for a reduced period of incubation. Enzymatic pretreatment of the DNA with a methylase will also generate a partial restriction digest. A uniform partial restriction digestion of DNA contained in agarose may, however, be difficult to obtain using the above mentioned methods since the diffusion rate of the restriction enzyme molecules is slow as compared to the speed of digestion itself. In order to obtain homogeneous partial digests of DNA in agarose plugs we devised a protocol where the DNA is incubated together with the enzyme but in absence of Mg^{2+}, the cofactor necessary for class II restriction endonuclease activity [1, 2]. The partial digestion is controlled by adding limiting amounts of Mg^{2+} to the plugs and allowing for 1 h incubation at 37 °C. This method is an easy and robust way of obtaining good partial restriction digests with Eco RI. A small sample of plugs from each preparation of DNA is tested in reactions with serial dilutions of Mg^{2+} in order to identify the Mg^{2+} concentration yielding the desired fragment size.

Protocol for partial EcoRI restriction digest
1. Plugs are rinsed three times 30 min in 50 volumes of restriction buffer lacking Mg^{2+} (100 mM NaCl, 10 mM Tris-HCl, pH 7.9).
2. The plugs are incubated in one volume of restriction buffer lacking Mg^{2+} with 5 u/µg of enzyme at 4 °C for 1 h.
3. Mg^{2+} from a stock solution of 100 mM $MgCl_2$ is added to the desired concentration to initiate the enzymatic reaction. (Or a series of reactions are set up with concentrations ranging from 0.01 mM to 1.00 mM).
4. The reactions are immediately transferred to 37 °C and incubated for 1 h.
5. The reactions are stopped by discarding the restriction buffer and adding a large excess of cold $TE_{10.5}$.

Complete restriction digestion

Complete restriction digestions can easily be made with DNA contained in agarose plugs and may serve (e.g.) in fingerprint analysis.

Protocol for complete restriction digest
1. Plugs are rinsed three times 30 min in 50 volumes of the appropriate restriction buffer.
2. The plugs are incubated in one volume of restriction buffer in the presence of 7–10 u/µg of enzyme at 37 °C (depending on the enzyme) for 3h.
3. The reactions are stopped by discarding the restriction buffer and adding a large excess of cold $TE_{10.5}$. The plugs are loaded onto an electrophoresis gel.

Preparation of the pYAC4 vector

The most commonly used yeast artificial chromosome vector is YAC4 which allows for cloning in a Eco RI site contained in the intron of the *SUP4* tRNA gene. The shuttle vector, produced in quantity in *E. coli*, is linearized with Bam HI that liberates a *HIS3* spacer fragment. In order to avoid ligation between the telomeres and the *HIS3* spacer, the Bam HI site is dephosphorylated. After phenol/chloroform extraction the cloning site is opened with Eco RI and following a second phenol/chloroform extraction the vector is ready to use.

149

Ligation in agarose

Large DNA molecules trapped inside an agarose plug essentially do not diffuse while small DNA molecules only do so at a slow rate. It is therefore necessary to briefly melt the plug prior to ligation in order to obtain a homogeneous mixture between vector molecules and insert molecules. The vector is added to the insert in a molar excess of 40 : 1 (this roughly means that the mass of vector is equal to the mass of insert). In order to obtain the maximum ligation efficiency the vector is only dephosphorylated at the telomeres (the *Bam* HI sites) and not at the cloning site (*Eco* RI site). This particular treatment further provides means for judging the ligation efficiency since the three YAC4 restriction fragments will ligate between themselves in a predictable way (RR : RL : LR : LL). The *Bam* HI spacer fragment carrying the *HIS3* gene will not ligate to anything. These ligation products as well as unligated substrate fragments are easily visualized in the CHEF size fractionation gels. This allows an estimation of the ligation efficiency (Fig. 2).

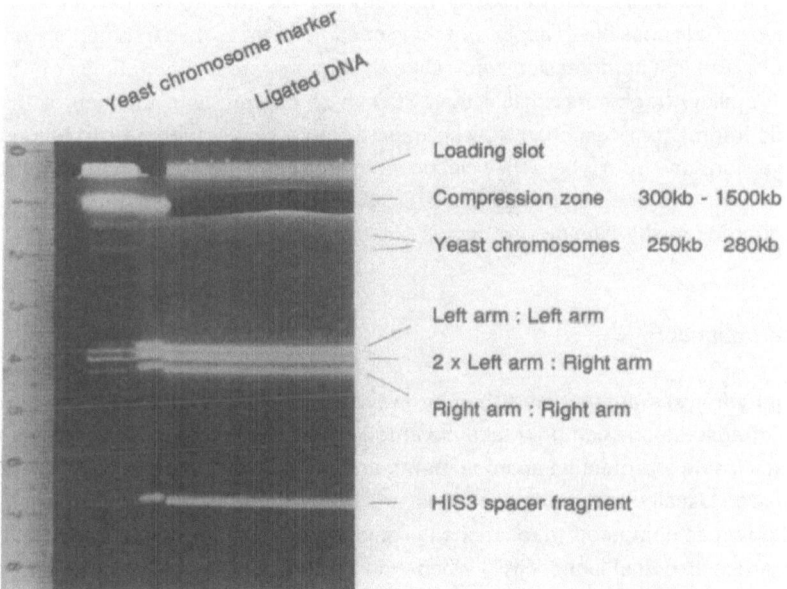

Fig. 2. Figure describing second CHEF sizing and ligation products. Following ligation the insert and vector DNA was loaded onto a CHEF gel and size-separated. The high-molecular-weight ligation products were retained in the compression zone and recovered in an agarose slice for transformation. Three bands corresponding to the three possible vector ligation products can be observed in the middle of the gel. The band at the bottom of the gel is the HIS3 spacer fragment that separates the telomeres in pYAC4. The efficiency of the ligation reaction can be assessed by the appearance of the vector ligation products and the disappearance of unligated vector arms.

Ligation in SeaPlaque agarose

1. 1 ml of SeaPlaque agarose recovered from the size fractionation gel is rinsed four times 30 min at room temperature as follows: first twice in 50 ml of bi-distilled water then twice in 1× ligation buffer.

2. The buffer is discarded and vector prepared for ligation is added to the plug in a quantity equal to the estimated weight of insert DNA (vector : insert molar ratio 40 : 1) and eventually the plug briefly melted in a 68 °C water bath and transferred to 37 °C.

3. 1 ml of 1× ligation buffer with 4000 units T4 ligase (Weiss units, New England Biolabs) dissolved and preheated to 37 °C is added and gently mixed to the liquid DNA. The ligation reaction is incubated at 37 °C for 2 h, then transferred to room temperature where the reaction is incubated overnight.

Size fractionation of DNA in CHEF gels

The transformation efficiency is affected by the size of the transforming YAC molecule, with small YACs transforming yeast spheroplasts more readily than larger molecules. This phenomenon will bias the representation of YACs in the library. It is therefore necessary to size-fractionate the DNA fragments so that only fragments larger than a particular threshold will be retained for transformation. In a 1% SeaPlaque gel it is possible to apply switching conditions that retain fragments above a particular size in a compression zone. On a CHEF apparatus, switch times of 15 s times 15 s allow fragments smaller than 300 kb to migrate as a function of their size while fragments larger than 300 kb migrate more slowly without resolution in a compression zone from where they can be recovered [2]. The electrophoresis buffer is 0.5× TBE, and the temperature is 10 °C. A schematic presentation of the individual steps in the cloning protocol is shown in Fig. 3.

Agarase treatment

Agarase hydrolyses agarose and thereby prevents polymerization. With the introduction of agarase one can now take advantage of allowing enzymatic reactions to take place inside a protective agarose matrix and liberate the processed DNA when appropriate. Usually enzymatic reactions are not inhibited by agarose itself but slowed down as compared to reactions in liquid. It is important to use low-melting-point agarose (like SeaPlaque, FMC) when working with DNA embedded in agarose. This is because the melting temperature of this type of agarose is slightly lower than the temperature at which DNA usually denatures. It is thus possible to take advantage of the agarose-hydrolysing enzyme agarase without denaturing the DNA itself [2, 12]. Agarose plugs containing YAC DNA can be hydrolysed with agarase and the solution transformed directly (in the presence of oligosaccharides) into yeast spheroplasts without further purification.

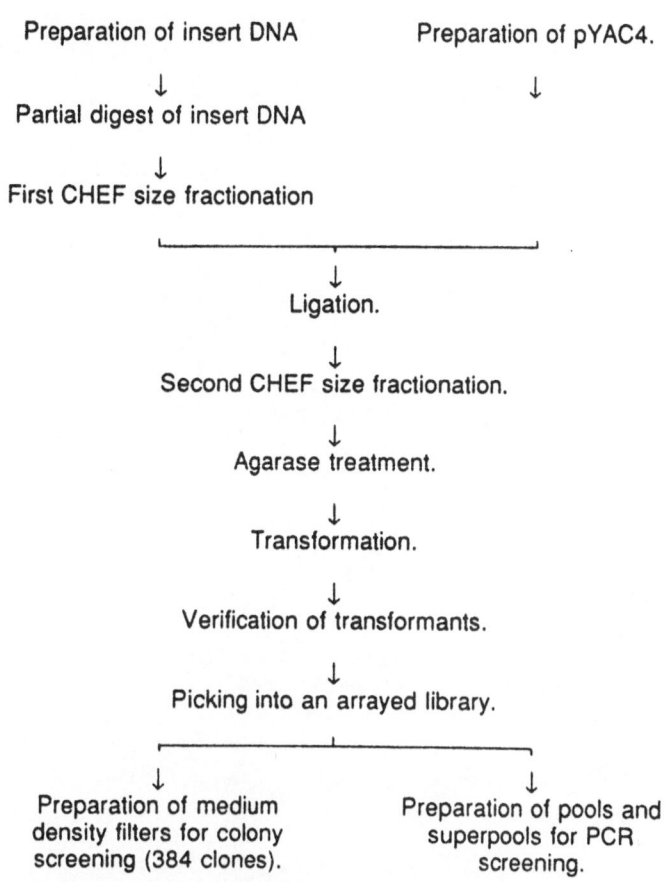

Fig. 3. Schematic presentation of the YAC cloning protocol.

Protocol for agarase treatment

1. A plug of agarose is equilibrated against 30 mM NaCl in distilled water. The solution is discarded and the mass of the plug is determined.
2. The plug is transferred to a 10 ml tube and incubated for 5 min in a 68 °C water-bath.
3. When the plug has melted the tube is transferred to 37 °C and agarase is added to 40–80 units per gram of agarose. The reaction is incubated for 2 h at 37 °C.

Transformation of spheroplasts

One of the limiting steps in YAC cloning is the low transformation efficiency in yeast. Transformation efficiencies are usually in the order of 3×10^3 YAC transformants per microgram of DNA. This low transformation efficiency is probably due to the particular morphology of yeast as well as to the large size of the transforming DNA. The most commonly used yeast transformation protocol is PEG-mediated transformation of spheroplasts described by Burgers and Percival [13]. A variety of enzymes including Zymolyase, lyticase and Novozyme 234 may be used to generate spheroplasts, but each has individual kinetics that must be determined in order to get satisfactory transformation efficiencies. We routinely use Zymolyase 20-T to prepare spheroplasts for YAC transformations. The protocol presented here is adapted to a transformation experiment with 50 Petri dishes. As it is difficult to control the spheroplasting process, we prepare four samples incubated with various concentrations of Zymolyase 20-T (1–20 units). The samples are prepared in parallel and only prior to transformation do we select which one of the four samples is to be used in the transformation experiment.

Spheroplast preparation
1. Inoculate 200 ml YPD with a single AB1380 colony and incubate at 30 °C with vigorous shaking. When the culture has reached mid-log growth phase, the cells are harvested in a centrifuge at 900 *g*.
2. The pellet is resuspended and rinsed twice in distilled water and once in 1 M sorbitol. The cells are counted under the microscope in a haemocytometer and four samples are aliquoted, each containing 1.5×10^9 cells. The four samples are harvested as above and resuspended in 5 ml SCEM.
3. Zymolyase 20-T (1–20 units/sample) is added to each sample and incubated with very gentle agitation at 30 °C for 10–15 min. (From this point on the cells are very fragile!).
4. The cells are harvested in a centrifuge at 500 *g* and the pellet washed once in 10 ml 1 M sorbitol. The cells are again harvested and the pellet resuspended in YPD buffer with 1 M sorbitol. The cells are left to recover for 30 min; then 5 ml of STC is added and the cells harvested. The pellet is rinsed once in 10 ml of STC and resuspended in 5 ml STC. The cells are at this point stable and ready for transformation.
5. The spheroplast preparations are analysed under phase-contrast microscope. The sample chosen for transformation shows no more that a few percent lysed cells in the STC buffer, but 100% lysed cells after water has been added to the microscope slide.

Transformation of spheroplasts
1. Agarose (0.5 mg) containing the ligated and size-fractionated DNA to be used for transformation is treated with agarase. Then 0.3 ml of 2 M sorbitol is added to the liquid DNA sample.
2. The DNA preparation is dispensed in 15 μl aliquots into 50 transparent 10 ml tubes.
3. 100 μl of the cell suspension is added and gently mixed to the reaction in each tube and incubated for 10 min at room temperature.
4. One ml of PEG is added and gently mixed to the reaction in each tube and incubated for 10 min. The cells are recovered in a centrifuge at 500 *g* and the supernatant thoroughly aspired without disturbing the pellet.
5. The pellets are resuspended in 150 μl of SOS and incubated at 30 °C for 45 min.
6. 3 ml of TOP lacking uracil warmed to 40 °C is added to each tube and transferred to Petri dishes containing SORB lacking uracil.
7. Transformants appear after 3—4 days of incubation at 30 °C.

Verification of YAC transformants

The primary transformation medium lacks uracil but not tryptophan. It is therefore recommended that all primary transformants are picked onto SD medium lacking both uracil and tryptophan. Clones that grow on this medium and have the red phenotype characteristic of the interrupted SUP4 tRNA gene comply with the genetic criteria of a YAC clone. A representative sample of clones from each transformation should also be tested on PFGE and by hybridization of Southern blots. Only when the tested clones are of the desired size (above the selected sizing threshold), are the transformants to be included in the expanding library.

Storage of the YAC library

An important issue for long-term storage and application of YAC libraries is whether YAC clones are maintained individually in ordered arrays in microtitre dishes or whether many YACs are stored mixed in pools. Arrayed libraries are ideal for large-scale physical mapping purposes whereas pooled libraries will work fine if the purpose is only to identify one or a few YACs of interest. It must be emphasized that replica plating and preparation of screening filters from arrayed libraries is very tedious and resource-demanding.

Protocol for triplicate storage of arrayed YAC libraries
This protocol describes the storage of a YAC library as one master library and two working libraries. The master library is only accessed in the case where live clones from given positions are present in neither of the two working libraries.
1. Inoculate single colonies in 600 μl of YPD in Micronics racks (96 × 1 ml) and incubate for 36 h with agitation at 30 °C. Inoculate the cultures in SD medium for colony filters and PCR pools.

2. Add 200 µl of 80% glycerol to each culture and mix well.
3. Transfer 200 µl of the cell suspension to each of three microtitre plates. Wrap in Saran wrap and store rapidly at −80 °C.
4. Cap the Micronics and store at −80 °C.

Protocol for non-organized storage of a YAC library
About 500 colonies are stored pooled in aliquots in 20% glycerol at −80 °C, and a preparation of DNA from the pooled culture is extracted for PCR assay. If a PCR pool is found positive then cells from the aliquot corresponding to the positive pool are plated on a hybridization filter on a Petri dish for colony screening.
1. After transformation, test for growth on SD medium lacking uracil and tryptophan.
2. Scrape off and pool 500 colonies that have grown on SD medium lacking both uracil and tryptophan. Use the cells to inoculate 500 ml of YPD and incubate at 30 °C with agitation for 6−8 h.
3. Take out 7×10^8 yeast cells to make 10 µg total DNA in 1 ml (enough for 500 PCR reactions).
4. Centrifuge the remaining part of the culture, resuspend pellet in 20% glycerol and store the pooled yeast in aliquots at −80 °C.
5. Inoculate about 2500 clones from an aliquot on SD medium lacking uracil and incubate for 48 h at 30 °C. Five replicas are made on nylon filters. Two of the colony filters are stored on Whatman paper soaked with 20% glycerol at −80 °C. The remaining filters are prepared for colony screening.

Screening of a YAC library

Several methods have been devised for screening YAC libraries [11, 28, 30, 31, 49]. They are all to some degree related, and have served to develop the protocol described here. The screening procedure is separated into three steps. The first step is PCR screening of pools of YACs. The virtue of the PCR screening is that it allows an easy way for exclusion of a large part of the library. This facilitates the second step, the colony screening on filters, so that it can be carried out on that particular sub-set of filters known to contain a positive colony (Fig. 4). Finally, a verification of the positive colonies is carried out. Alkaline transfer and radiolabelling of DNA can be performed as described [44].

PCR screening of pooled YACs

The strategy for screening pools of YAC clones by PCR is derived from Green and Olson [26], and the method, previously described, for extraction of YAC DNA is from Anand *et al*. [4]. The PCR screening is carried out in two steps. In the first step superpools of 1920 colonies are screened (corresponding to twenty 96-well microplates). The second step is screening of the five pools of 384 colonies (corresponding to four 96-well microplates) included in the positive superpools.

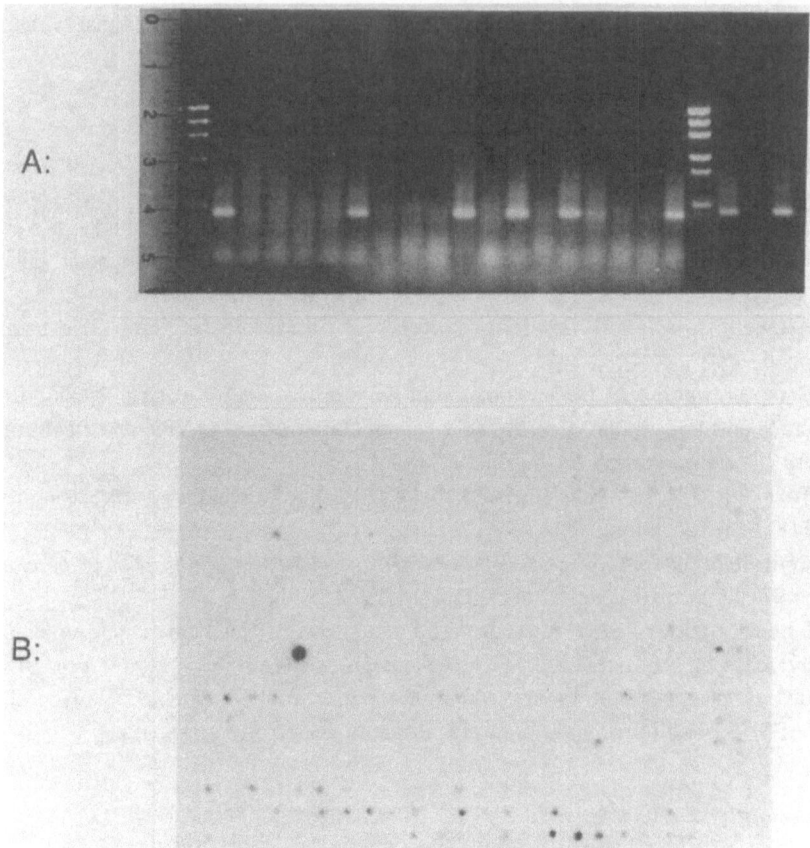

Fig. 4. PCR and colony screening. The presence or absence of a particular locus in a given pool of YAC DNA can be tested using either PCR or colony hybridization. A. If a pool is positive the resulting PCR product can be visualized on an agarose gel. Six positive pools are observed between the two DNA size markers. The three right most lanes are positive and negative controls. B. Yeast colonies were grown on a 11 cm by 7 cm nylon membrane and prepared for colony hybridization. Following hybridization a single strong signal indicated the position of the positive YAC colony. A background of unspecific hybridization helps identifying the correct coordinates of the clone.

PCR screening protocol

PCR reactions are performed with 20 ng of DNA from a yeast superpool in a final volume of 10 µl overlayered with a drop of light mineral oil. The PCR cycle is dependent on the primer set as well as the length of the PCR product. PCR products analysed on 1.5% SeaKem gels, or for very small fragments on gels of 3% SeaKem and 1% NuSieve.

Screening of YACs by colony hybridization

Preparation of filters for colony screening

1. Thaw the glycerol stocks and inoculate in 150 µl SD medium lacking uracil in microtitre plates at 30 °C for 3 days. These cultures are used for inoculation of membranes and to prepare YAC DNA pools.
2. Inoculate a charged nylon membrane (Biodyne B, Pall) with 384 YAC clones from these cultures. We used a replicator with 96 stainless steel pins arrayed in the microtitre plate configuration.
3. Place the inoculated membranes on SD medium lacking uracil and tryptophan on 22.5 × 22.5 cm plates (Nunc) and incubate at 30 °C for 2–3 days until colonies are about 2 mm in diameter (too large colonies are difficult to lyse). Avoid any bubbles under the membranes at all steps.
4. Remove the membranes and put them on the lid of the plates. Throw away the agar and place in the plate a 22 × 22 cm 3MM Whatman paper soaked in SCEM containing 0.5 U/ml Zymolyase 20-T. Place the membranes on the saturated paper. Seal the Nunc plates with parafilm and incubate at 30 °C overnight.
5. Incubate the membranes at room temperature on 3MM Whatman saturated with:
   - 10% SDS   5 min
   - 0.5 M NaOH   10 min
   - transfer to dry Whatman 3MM   5 min
   - 0.2 M Tris-HCl pH 7.5, 2× SSC   5 min 3 times
   - air-dry   2 hours
6. Store the dried membranes at 4 °C. The membranes can also be stored in pre-hybridization buffer.

Hybridization protocol

1. Pre-hybridize the membranes for at least 3 h at 65 °C in Tupperware boxes (20 ml + 1 ml/filter).
2. Pre-hybridization and hybridization buffers contain 7% PEG 8000 and 10% SDS. 100 µg/ml of sonicated denatured salmon sperm DNA is added to the pre-hybridization buffer.
3. Hybridization is carried at 65 °C with at least 3×10^5 cpm/ml.
4. Washes are carried out at stringencies particular to each probe.

Filters can be hybridized at least 10 times without significant loss of signal (do not dry membranes once they have been hybridized). Usually overnight exposure give good signals.

Identification of a positive YAC clone

Filter screening will, from time to time, give ambiguous answers as to which clone in the grid is truly positive. A final verification of all selected clones should therefore be carried out.

Protocol for verification of a positive YAC

1. The YAC clone from the glycerol stock is streaked out on SD medium lacking uracil and incubated at 30 °C for 3 days.
2. From a single colony a 5 ml culture in YPD is inoculated and incubated at 30 °C with agitation overnight.
3. The cells are harvested and 2 agarose plugs of 100 μl prepared and chromosomes purified using the protocols previously described.
4. The chromosomes are size-separated on PFGE and Southern blots of the gels are hybridized in order to determine the insert size of the positive YACs.
5. The final step is to prepare 20% glycerol stocks of the positive YAC colonies and store them at −80 °C.

YAC clones in physical mapping

Several strategies can be applied to either link a YAC to a contig or to extract a refined physical map from a YAC. These manipulations, even though they appear straightforward, involve very large DNA fragments and will often reveal very complex information that can be difficult to interpret.

For chromosome walking experiments, two methods have been used to isolate the extremities from the DNA insert. The first, plasmid rescue, takes advantage of the genetic markers in the YAC vector that may be selected for in bacteria. Unfortunately, with pYAC4, only the extremity next to the centromere can be rescued in this way. The second method is based on a PCR assay and several variants hereof have been described. One is designed so that one primer recognizes the vector sequence next to the cloning site and the other primer recognizes a highly repeated sequence specific to the source genome [9, 32, 36]. Another method uses a specialized adapter, the Vectorette [42], that is ligated to the restriction-digested insert. Due to its design and the choice of primers, only fragments containing part of the YAC vector (first primer) and the Vectorette (second primer) will give rise to a PCR product. Alternatively, one may try inverse PCR [40, 48]. This is a technique, where the YAC is digested with a frequently cutting restriction enzyme, and individual fragments circularized. Junction fragments containing the cloning sites and a part of both vector and insert can be amplified with PCR using amplimers recognizing the vector.

Growth media and buffers

− YPD
  − 1% yeast extract (Difco)
  − 2% bactopeptone (Difco)
  − 2% D-glucose (Sigma G-8270)
  − 1.5−2% bactoagar (Difco)
  Autoclave, pH 5.8

- SCEM
  - 1 M sorbitol (Sigma S-1876)
  - 10 mM EDTA pH 8
  - 0.1 M Na citrate pH 5.8
  Autoclave and add 30 mM 2-mercaptoethanol

- STC
  - 1 M sorbitol
  - 10 mM Tris-HCl pH 8
  - 10 mM $CaCl_2$
  Autoclave

- PEG
  - 20% PEG 8000 (Sigma P-5413)
  - 10 mM Tris-HCl pH 8
  - 10 mM $CaCl_2$
  Filter sterilize

- SOS
  - 1 M sorbitol
  - 6.5 mM $CaCl_2$
  - 0.25% yeast extract (Difco)
  - 0.5% bactopeptone
  - 20 μg/ml Uracil and tryptophan
  Filter sterilize, pH 5.8

- SORB
  - 0.9 M sorbitol
  - 3% D-glucose
  - 0.67% Yeast Nitrogen Base without amino acids
  - 1.5–2% bactoagar
  Autoclave, pH 5.8
  Add amino acids as required

- TOP
  - 1 M sorbitol
  - 0.67% Yeast Nitrogen Base without amino acids
  - 2% D-glucose
  - 1% bactoagar
  Autoclave, pH 5.8
  Add amino acids as required

- SD
  - 0.67% Yeast Nitrogen Base without amino acids
  - 2% D-glucose
  Autoclave and add amino acids as required

– Zymolyase 20-T (Selkagaku Kogyo 120491):
  – 1000 U/ml in redistilled water.
  Filter sterilize and store at 4 °C.

– Agarase (Calbiochem 121814):
  – 2000 U/ml in redistilled water with 50% glycerol.
  Store at −20 °C.

Amino acids:

Sigma	URA-	URA- TRP-
Ade A 9795	1	1
Arg A 3909	4	–
His H 9511	2	2
Iso I 7383	6	6
Leu L 1512	6	6
Lys L 1262	5	5
Met M 2893	2	2
Phe P 5030	5	5
Thr T 1645	20	20
Trp T 0271	4	–
Tyr T 1020	5	5
Total:	60	52
use at:	600 mg/l	520 mg/l

Application of a YAC library

Storage of information

It is our experience that the data generated through application of a YAC library quickly becomes difficult to handle manually. One answer to this problem is the use of a computerized database, where general information of the library as well as specific data from individual clones and probes screened on the library can be stored.

Stability of YACs

YAC clones are generally stable over many generations. The artificial chromosomes segregate like the natural chromosomes during mitosis and are present in a single copy in each yeast cell. Occasionally YAC clones are observed that appear to diverge from this pattern. Apparently three problems, that seem to vary in importance with different YAC libraries, tend to obscure genomic origin. The first is co-ligation events, where two or more independent DNA segments are cloned into one YAC. Secondly, two or more YACs may be transformed into one yeast cell. The third problem concerns the clonal stability of YACs. Each of these problems may be analysed to a certain extent. The extent of the first problem could be determined by analysing a YAC library made from a mixture of DNA from two different organisms. The fraction of clones containing DNA from both sources can be determined by colony screening. Co-transformation events are identified during the verification process and appear as two distinct bands on Southern blots. As for clonal instability of YACs, we and others have occasionally observed YACs that appear as multiple bands on Southern blots. Analysis of subclones shows that each carries a different single or subset of the bands of the original YAC. Neil *et al.* [37] found that by using recombination-deficient yeast strains, rad52 strains in particular, the recombination events are significantly reduced.

YACs – a commitment

The construction of a YAC library is very time- and labour-consuming, as is the exploitation of the library. Therefore, careful considerations should be made of which species and genotype would be particularly interesting to use as DNA source, and whether the clones should be stored in an arrayed library or with a lesser level of organization. It is equally important to consider how to share the library with other investigators.

Other applications of YACs

In this chapter we have primarily described the construction of YACs and their application in physical mapping. YACs, however, have further proved their biological usefulness in transfection experiments where YACs introduced into mammalian cell lines have complemented enzymatic deficiencies. Biological applications of YACs in plant genetics will be very interesting to investigate and might become a powerful new tool in plant breeding.

Acknowledgements

We would like to acknowledge G. Carle for advice and discussions, and D. Burke and M. Olson for the YAC vectors and the yeast strain AB1380.

H.M.A. is supported by a fellowship from the Danish Medical Research Council.

163

References

1. Albertsen HM, Le Paslier D, Abderrahim H, Cann H, Dausset J, Cohen D (1989) Improved control of partial DNA restriction enzyme digest in agarose using limiting concentrations of Mg^{++}. Nucleic Acids Res 17: 808.
2. Albertsen HM, Abderrahim H, Cann H, Dausset J, Le Paslier D, Cohen D (1990) Construction and characterization of a yeast artificial chromosome library containing seven haploid human genome equivalents. Proc Natl Acad Sci USA 87: 4256–4260.
3. Anand R, Villasante A, Tyler-Smith C (1989) Construction of yeast artificial chromosome libraries with large inserts using fractionation by pulsed-field gel electrophoresis. Nucleic Acids Res 17: 3425–3433.
4. Anand R, Riley JH, Smith JC, Markham AF (1990) A 3.5 genome equivalent multi access YAC library: construction, characterisation, screening and storage. Nucleic Acids Res. 18: 1951–1955.
5. Bates GP, MacDonald ME, Baxendale S, Sedlacek Z, Youngman S, Romano D, Whaley WL, Allitto BA, Poustka A, Gusella JF, Lehrach H (1990) A yeast artificial chromosome telomere clone spanning a possible location of the Huntington Disease gene. Am J Hum Genet 46: 762–775.
6. Beckmann J, Soller M (1986) Restriction fragment length polymorphisms in plant genetic improvement. Oxford Surv Plant Mol Biol Cell Biol 3: 196–250.
7. Bellis M, Pages M, Roizes GA (1987) A simple and rapid method for preparing yeast chromosomes for pulsed field gel electrophoresis. Nucleic Acids Res 15: 6749.
8. Botstein D, White RL, Skolnick M, Davis RW (1980) Construction of a genetic linkage map in man using restriction fragment length polymorphisms. Am J Hum Genet 32: 314–331.
9. Breukel C, Wijnen J, Tops C, van der Klift H, Dauwerse H, Meera Khan P (1990) Vector-Alu PCR: a rapid step in mapping cosmids and YACs. Nucleic Acids Res 18: 3097.
10. Brown WRA (1989) Molecular cloning of human telomeres in yeast. Nature 338: 774–776.
11. Brownstein BH, Silverman GA, Little RD, Burke DT, Korsmeyer SJ, Schlessinger D, Olson MV (1989) Isolation of single-copy human genes from a library of yeast artificial-chromosome clones. Science 244: 1348–1351.
12. Bucan M, Yang-Feng T, Colbert-Poley AM, Wolgemuth D, Guenet JL, Francke U, Lehrach H (1986) Genetic and cytogenetic localisation of the homeobox containing genes of mouse chromosome 6. EMBO J 5: 287–290.
13. Burgers PMJ, Percival KJ (1987) Transformation of yeast spheroplasts without cell fusion. Anal Biochem 163: 391–397.
14. Burke DT, Carle GF, Olson MV (1987) Cloning of large segments of exogenous DNA into yeast by means of artificial chromosome vectors. Science 236: 806–812.
15. Carle GF, Olson MV (1984) Separation of chromosomal DNA molecules from yeast by orthogonal-field alternation gel electrophoresis. Nucleic Acids Res 12: 5647–5664.
16. Carle GF, Frank M, Olson MV (1986) Electrophoretic separations of large DNA molecules by periodic inversion of the electric field. Science 232: 65–68.
17. Cheng J-F, Smith CL, Cantor CR (1989) Isolation and characterization of a human telomere. Nucleic Acids Res 17: 6109–6127.
18. Cheung WY, Gale MD (1990) The isolation of high molecular weight DNA from wheat, barley and rye for analysis by pulsed-field gel electrophoresis. Plant Mol Biol 14: 881–888.
19. Chu G, Vollrath D, Davis RW (1986) Separation of large DNA molecules by contour-clamped homogenous electric fields. Science 234: 1582–1585.
20. Clarke L, Carbon J (1976) A colony bank containing synthetic Col E1 hybrid plasmids representative of the entire E. coli genome. Cell 9: 91.
21. Collins FS, Weissman SM (1984) Directional cloning of DNA fragments at a large distance from an initial probe: A circularization method. Proc Natl Acad Sci USA 81: 6812–6816.
22. Coulson A, Waterson R, Kiff J, Sulston J, Kohara Y (1988) Genome linking with yeast artificial chromosomes. Nature 335: 184–186.
23. Cross SH, Allshire RC, McKay SJ, McGill NI, Cooke HJ (1989) Cloning of human telomeres by complementation in yeast. Nature 338: 771–774.

164

24. Feil R, Palmieri G, d'Urso M, Heilig R, Oberle I, Mandel JL (1990) Physical and genetic mapping of polymorphic loci in Xq28 (DXS15, DXS52, and DXS134): Analysis of a cosmid clone and a yeast artificial chromosome. Am J Hum Genet 46: 720–728.
25. Garza D, Ajioka JW, Burke DT, Hartl DL (1989) Mapping the *Drosophila* genome with yeast artificial chromosomes. Science 246: 641–646.
26. Green ED, Olson MV (1990) Systematic screening of yeast artificial-chromosome libraries by use of the polymerase chain reaction. Proc Natl Acad Sci USA 87: 1213–1217.
27. Guzman P, Ecker JR (1988) Development of large DNA methods for plants: molecular cloning of large segments of *Arabidopsis* and carrot DNA into yeast. Nucleic Acids Res 16: 11091–11105.
28. Heard E, Davies B, Feo S, Fried M (1989) An improved method for the screening of YAC libraries. Nucleic Acids Res 17: 5861.
29. Kuspa A, Vollrath D, Cheng Y, Kaiser D (1989) Physical mapping of the *Myxococcus xanthus* genome by random cloning in yeast artificial chromosome. Proc Natl Acad Sci USA 86: 8917–8921.
30. Labella T, Schlessinger D (1989) Complete human rDNA repeat units isolated in yeast artificial chromosomes. Genomics 5: 752–760.
31. Lai E, Cantrell C (1989) Rapid colony screening of YAC libraries by using alginate as matrix support. Nucleic Acids Res 17: 8008.
32. Ledbetter SA, Nelson DL, Warren ST, Ledbetter DH (1990) Rapid isolation of DNA probes within specific chromosome regions by interspersed repetitive sequence (IRS) PCR. Genomics 6: 475–481.
33. Little RD, Porta G, Carle GF, Schlessinger D, D'Urso M (1989) Yeast artificial chromosomes with 200- to 800-kilobase inserts of human DNA containing HLA, V_K, 5 S, and Xq24-Xq28 sequences. Proc Natl Acad Sci USA 86: 1598–1602.
34. Murray AW, Schultes NP, Szostak JW (1986) Chromosome length controls mitotic chromosome segregation in yeast. Cell 45: 529–536.
35. Murray AW, Szostak JW (1983) Construction of artificial chromosomes in yeast. Nature 305: 189–193.
36. Nelson DL, Ledbetter SA, Corbo L, Victoria MF, Ramirez-Solis R, Webster TD, Ledbetter DH, Caskey CT (1989) Alu PCR: a method for rapid isolation of human-specific sequences from complex DNA sources. Proc Natl Acad Sci USA 86: 6686–6690.
37. Neil DL, Villasante A, Fisher RB, Vetrie D, Cox B, Tyler-Smith C (1990) Structural instability of human tandemly repeated DNA sequences cloned in yeast artificial chromosome vectors. Nucleic Acids Res 18: 1421–1428.
38. Pavan WJ, Hieter P, Reeves RH (1990) Generation of deletion derivatives by targeted transformation of human-derived yeast artificial chromosomes. Proc Natl Acad Sci USA 87: 1300–1304.
39. Poustka A, Pohl TM, Barlow DP, Frischauf A-M, Lehrach H (1987) Construction and use of human chromosome jumping libraries from NotI-digested DNA. Nature 325: 353–355.
40. Ochman H, Gerler AS, Hartl DL (1988) Genetic applications of an inverse polymerase chain reaction. Genetics 120: 621–623.
41. Riethman HC, Moyzis RK, Meyne J, Burke DT, Olson MV (1989) Cloning human telomeric DNA fragments into *Saccharomyces cerevisiae* using a yeast-artificial-chromosome vector. Proc Natl Acad Sci USA 86: 6240–6244.
42. Riley J, Butler R, Ogilvie G, Jenner D, Powell S, Anand R, Smith JC, Markham AF (1990) A novel, rapid method for the isolation of terminal sequences from yeast artificial chromosome (YAC) clones. Nucleic Acids Res 18: 2887–2890.
43. Saiki RK, Scharf S, Faloona F, Mullis KB, Horn GT, Erlich HA, Arnheim N (1985) Enzymatic amplification of β-globin genomic sequences and restriction site analysis for diagnosis of sickle cell anemia. Science 230: 1350–1354.
44. Sambrook J, Fritsch EF, Maniatis T (1989) Molecular Cloning: A Laboratory Manual (2nd ed.) Cold Spring Harbor, NY: Cold Spring Harbor Laboratory Press.
45. Schwartz DC, Cantor CR (1984) Separation of yeast chromosome-sized DNAs by pulsed field gradient gel electrophoresis. Cell 37: 67–75.

46. Sheeman C, Weiss AS (1990) Yeast artificial chromosomes: rapid extraction for high resolution analysis. Nucleic Acids Res 18: 2193.
47. Sherman F, Fink GR, Hicks JB (1986) Methods in Yeast Genetics. Cold Spring Harbor, NY: Cold Spring Harbor Laboratory.
48. Silverman GA, Ye RD, Pollock KM, Sadler JE, Korsmeyer SD (1989) Use of yeast artificial chromosome clones for mapping and walking within human chromosome segment 18q21.3. Proc Natl Acad Sci USA 86: 7485–7489.
49. Traver CN, Klapholz S, Hyman RW, Davis RW (1989) Rapid screening of a human genomic library in yeast artificial chromosomes for single-copy sequences. Proc Natl Acad Sci USA 86: 5898–5902.
50. Wada M, Little RD, Abidi F, Porta G, Labella T, Cooper T, Della Valle G, D'Urso M, Schlessinger D (1990) Human Xq24-Xq28: Approaches to mapping with yeast artificial chromosomes. Am J Hum Genet 46: 95–105.
51. Ward ER, Jen GC (1990) Isolation of single-copy-sequence clones from a yeast artificial chromosome library of randomly-sheared *Arabidopsis thaliana* DNA. Plant Mol Biol 14: 561–568.

8. Linkage analysis in human genetics

JEAN-MARC LALOUEL

Howard Hughes Medical Institute, Eccles Institute of Human Genetics, University of Utah, Salt Lake City, UT 84132, USA

Introduction

Linkage analysis in humans owes its significance to the opportunity it offers to delineate the genetic basis of clinical disorders, and to the avenues it opens for the eventual identification of elusive molecular defects. Although different experimental designs are used for linkage analysis in plants, the same basic principles are involved. A brief account of the experience accumulated with linkage studies in humans might be helpful for designing experiments in plant or animal sciences. For in-depth reviews of the topic, the reader is referred to White and Lalouel [36] or to textbooks such as Bailey [1] and Ott [28].

Experimental design

It would seem that lack of experimental control should, at the outset, have been enough of a deterrent to limit the practical feasibility of linkage analysis in humans. Early work with laboratory animals led to the development of simple, yet powerful breeding experiments to yield tests of linkage or estimates of recombination through statistical methods involving familiar test statistics and a limited amount of hand calculations, as reviewed in Bailey [1]. The double backcross design (AaBb × AABB) can serve as a prototype (Table 1). The distribution of alleles on the parental chromosomes, or phase, in the double heterozygote is determined by the experimental crosses performed, in either one of two possible configurations traditionally referred to as 'coupling' (AB/ab) or 'repulsion' (Ab/ab). It follows that recombination events can be identified unambiguously in the progeny. Hence the frequency of recombination can be calculated directly as the proportion of observed crossover progeny, and simple chi-square tests can be used to detect significant linkage.

In man, these designs can be approached only by selecting and testing families until informative matings are identified. In a nuclear family consisting of two parents and several offspring, the situation analogous to a double backcross does not define the phase of the alleles at both loci; that is, the double heterozygote AaBb can be in either configuration, AB/ab or Ab/aB. This fact alone accounts for the greatest departure of linkage strategies in humans from those applied to experimental species. As illustrated in Table 1, recombination

167

J.S. Beckmann and T.C. Osborn (eds) Plant Genomes: Methods for Genetic and Physical Mapping, 167–180.

Table 1. Detection of linkage in a double backcross, $AaBb \times aabb$

Phase of first parent	Prior probability			
Coupling: $AB\|ab$	$1\|2$			
Repulsion: $Ab\|aB$	$1\|2$			

Conditional probabilities of gametes produced by first parent

Phase	Games: AB	Ab	aB	ab	Total
Coupling	$(1-r)\|2$	$r\|2$	$r\|2$	$(1-r)\|2$	1
Repulsion	$r\|2$	$(1-r)\|2$	$(1-r)\|2$	$r\|2$	1

Observed progeny (O)

Progeny genotype	$AB\|ab$	$Ab\|ab$	$aB\|ab$	$ab\|ab$	Total
Observed number	n_1	n_2	n_3	n_4	n

$$p(O\,|\,r) = p(\text{coupling})p(O\,|\,\text{coupling},\, r) + p(\text{repulsion})p(O\,|\,\text{repulsion},\, r)$$
$$= 0.5[0.5^n r^{n2+n3}(1-r)^{n1+n4}] + 0.5[0.5^n r^{n1+n4}(1-r)^{n2+n3}]$$
$$p(O\,|\,0.5) = 0.25^n = 0.5^{2n}$$
$$z(r) = \log[p(O\,|\,r)/p(O\,|\,0.5)]$$
$$= \log 2^{n-1}[r^{n2+n3}(1-r)^{n1+n4} + r^{n1+n4}(1-r)^{n2+n3}]$$

events cannot be detected directly by identification of crossover progeny. Rather, a probabilistic argument is constructed, which expresses the probability of the observed progeny conditional on two equally probable parental phases. Recombination is then estimated as that value which maximizes the probability of the observations, following the likelihood principle which states that the best-supported hypothesis is that for which the probability of the data is maximal.

The experimental design of a human linkage study is dictated by the nature of the traits to be investigated. When the objective is to identify the genetic location of an unknown gene that accounts for an inherited disorder, the experimenter must rely on the ascertainment of families with multiple affected individuals. These can be nuclear families, or they may be extended pedigrees. The latter are commonly sampled for dominantly inherited, rare Mendelian conditions: having ascertained a proband, or index case presenting with a given disorder, the most effective way for the investigator to collect data on a substantial number of affected individuals consists of a systematic investigation of all relatives of the proband who can be examined. Hence, the structures of the families sampled is primarily dictated by nature.

Greater experimental flexibility is afforded when genotypes are generated in the laboratory for the specific purpose of constructing genetic maps. The design that has prevailed in humans consists of a collection of 'nuclear pedigrees', each including a large number of offspring, both parents, and whenever feasible all four grandparents [35]. Grandparents provide an internal control on the definition of segregating alleles in families, and their genotypes may often determine the parental phase of alleles in one or both parents.

This design defined a panel of reference families who have contributed the cell lines used by more than 60 laboratories to construct genetic maps of human chromosomes in an international collaboration [5]. Genetic markers are characterized on DNA produced from lymphoblastoid cell lines, and data are shared by all laboratories. This successful collaboration may serve as a model for investigators who intend to construct genetic maps of various plant or animal species.

Statistical inference

As we mentioned, tests of linkage in controlled crosses often require no more that a chi-square test of departure of observed numbers of crossover progeny from expected proportions under independent segregation, and estimation of recombination is usually straightforward. More elaborate models have been constructed to account for factors such as differential viabilities [1], but even such instances have involved simple statistics.

Tests of linkage in humans require computations of probabilities, and likelihood-ratio tests that contrast two competing hypotheses, namely linkage versus independent segregation. These calculations involve elementary probability distributions to define the prior probability distribution of alleles at each locus, the conditional probability of the genotype of an offspring given the genotypes of the parents, and the probability of an individual's phenotype given his genotype. When extended pedigrees are considered, hand-calculation becomes impractical. Rather, recursive algorithms are applied with the help of computer programs such as LIPED [26], LINKAGE [14] or MAPMAKER [12].

The statistical tests and assessment of statistical significance that have prevailed in human linkage studies depart somewhat from standard likelihood theory. Rather than relying on the asymptotic chi-square distribution of the likelihood-ratio test, so commonly used in statistics, one applies decimal logarithms to probability ratios, which are then referred to as 'LOD scores', where LOD stands for 'logarithm of the odds'. Linkage in humans is usually inferred when the maximum LOD score, as a function of the recombination fraction, exceeds the conventional value of 3.0. The historical basis of this unusual approach stems from a seminal paper [20] where the test was derived within the framework of sequential analysis [33]. Morton noted that linkage tests would often fit into a sequential sampling framework as more and more families were inspected with regard to a particular linkage, often through the cumulation of data arising in different laboratories. To allow an easy combination of data across distinct published reports without having to recompute probabilities from pooled raw data, Morton proposed that LOD scores be reported for a series of fixed, conventional values of the recombination fraction. For each value, LOD scores could be added across studies to assess significance of linkage. The convenience of LOD tables for reporting linkage was so overwhelming that LOD scores have become the acknowledged standard for reporting and testing the significance of linkage, even in the absence of a sequential sampling frame.

Tests of significance under this approach, however, and in particular the accepted value of 3.0 to declare significance, will continue to be shrouded in confusion both for casual users of the test and for statisticians. A number of elements further complicate the issue: the prior probability that two loci belong on the same chromosome, the common use of multiple genetic markers, the estimation of sex-specific recombination, and the joint estimation of parameters of a genetic model have each received some statistical treatment. Although a fair attempt to cover these issues can be found in a recent text [28], most investigators remain perplexed by the unusual treatment that the problem of detection of linkage has received in the human genetic literature. The confusion is compounded by the fact that standard likelihood ratio tests are used to test heterogeneity or sex difference in recombination.

Genetic models

Whenever linkage is sought between a clinical condition and one or more genetic markers, one must specify a model of the genetic determination of the clinical phenotype. This is straightforward when penetrance is complete and single-gene inheritance is postulated. Any departure from such basic models will require that parameters be specified to account for the incomplete determination of phenotype by genotype.

For discrete traits, such as the occurrence of disease, parameters are usually specified through the concept of incomplete penetrance. A single gene, autosomal or sex-linked, is postulated and penetrances specify the conditional probabilities of disease given each genotype. This formula will reflect dominance effects, incomplete determination, and delayed manifestation; the generalized single-locus model already encompasses multifactorial determination. It implies the assumption that a single gene exerts a major effect, and that although other factors contribute to phenotypic expression, the latter are either not inherited or their inheritance can be ignored.

The multifactorial model of inheritance, of the biometrical school, lies at the other end of the spectrum of genetic modeling. In such multifactorial or polygenic inheritance, it is assumed that phenotypes result from the cumulative effect of a large number of independent genetic and environmental effects. This very democratic form of genetic government was initially proposed for modeling the inheritance of continuous variables within the framework of the multivariate normal distributions during the early blossoming of multivariate distributions in the statistical sciences. These models have met with enormous success in animal and plant breeding, where they provide a rigorous framework for genetic improvement by controlled selection.

Multifactorial models came into vogue in the early 1960's in human genetics when it was perceived that conditions with low recurrence risks within families, as is the case for most congenital malformations, could result from the combined effects of several genetic and environmental determinants. The multifactorial model of quantitative inheritance was applied to discrete traits by defining the latter as threshold characters on an underlying continuous scale of liability to disease [6]. Such models do not further our understanding of causation of disease; indeed, they lack specificity, in that they do not put forward a hypothesis which further experiments can profitably attempt to refute. Rather, their merit, as in animal or plant breeding, lies in prediction; thus recurrence risks to relatives of various degrees can be expressed as a function of heritability estimates fitted to familial data.

Later, both single-gene and multifactorial models were combined in mixed models of inheritance, where a continuous variable is assumed to result from the cumulative effects of a major locus, multifactorial inheritance, and random environmental effects [7, 22]

Mixed models of inheritance afford proper nesting of hypotheses on a continuum encompassing Mendelian as well as multifactorial inheritance [11]. However, they are of much greater significance from a conceptual point of view. Indeed, they formally reconcile the contrasting approaches of the Mendelian and the biometrical schools, a logical dichotomy with historical rather than rational justification. Furthermore, such models also dissipate the parallel, and equally ill-justified, distinction between discrete and continous traits. It is common for human geneticists to analyze the segregation of quantitative traits in pedigrees under the assumption that a major gene contributes to the inheritance of the trait (for example, see [11]). This assumption holds also for linkage analysis, as can be seen in Lathrop et al. [13]. In fact, several quantitative traits have also been jointly considered. When linkage is sought, it is common to reduce the complexity of the genetic model by retaining only a major-gene effect and ignoring residual familial correlations.

The genetic analysis of quantitative traits does raise statistical issues regarding scale of measurement and the distribution of a quantitative trait, but this problem is not peculiar to genetic modeling. Whenever the distribution of a continuous variable departs from a normal distribution, a fundamental question is whether this departure results from the choice of the measurement scale or from an underlying major effect that yields a mixture of distributions. Simple transformations or families of power transformations can be considered. While this issue has received appropriate attention in human genetics [18], the resolution of skewness from commingling of distribution remains a difficult statistical challenge.

The relevant conclusions for animal and plant sciences are that genetic analyses of quantitative traits can be performed under a major-gene model, that there is no logical discontinuity between Mendelian and biometric approaches, and that no particular distinction is necessary between discrete and continuous traits.

Etiological heterogeneity

A lesson learned alike from the genetic analysis of inherited conditions and from biological measurements is that heterogeneity in causation is likely to exist for most phenotypes, and it can obscure or even lead to false inferences. Heterogeneity has been observed even for Mendelian conditions exhibiting very well defined phenotypes and complete penetrance; disorders of collagen biosynthesis, in particular the Ehlers-Danlos syndromes, can be referred to as an illustration [3].

There are a number of ways to handle this issue in the context of segregation and linkage analysis. One source of heterogeneity is the occurrence of pheno-copies which are not inherited but mimic the phenotype under study. Such sporadic events can be accounted for in a major-gene model by assuming that the probability that an individual of normal genotype presents the disease is not null, and that this probability can be estimated, even crudely. Were this precaution not considered, one conceivably could miss evidence of linkage.

When more than one gene can individually account for the observed pheno-type, heterogeneity is likely to occur among pedigrees. Linkage of a marker to one locus may not be detected if families segregating a defective allele at the other locus are included in the analysis. Two methods have been proposed to deal with this important issue. Morton [21] demonstrated by a likelihood-ratio test of heterogeneity among families that one form of elliptocytosis was linked to the red-cell polymorphism Rh while another, clinically indistinguishable, was not; this test admits $(n\text{-}1)$ degrees of freedom for n families. Another test [27, 31], postulates *a priori* that the sample of independently ascertained fami-lies consists of a mixture of 'linked' and 'unlinked' families. Likelihoods are expressed as a function of both recombination and the proportion of 'linked' families, yielding a likelihood-ratio test of heterogeneity with one degree of freedom. Of course, inferences drawn from a single, large pedigree would circumvent this difficulty, although clearly that strategy would not be adequate for common disorders.

Multilocus analysis and genetic mapping

Whenever more than two loci are linked, one can construct a genetic map by inferring their most likely order. Early in this century, detailed genetic maps were constructed for a number of laboratory species. Genetic mapping in *Drosophila* offers the best example of the strategy followed. For loose linkage, the very large number of progeny allowed investigators to determine gene order from pair-wise recombination data. Recognizing the greater power afforded by three-point or higher-order crosses, investigators designed multiple back-crosses with known phase in order to identify all recombinant progeny at several loci jointly. In a triple backcross with known phase, each gamete can be classified unambiguously as belonging to one of four possible outcomes, and three recombination rates can be directly estimated, r(A, B), r(B, C) and r(A, C); the two markers exhibiting the largest recombination rate are taken as flanking loci.

In humans, this situation cannot usually be achieved; the haphazard distribution of informative parental genotypes in a reference sample of families leads to incomplete data. Therefore tests of gene orders require likelihood calculations.

Recombination results from the occurrence of an odd number of crossing-over events and is not a proper metric for genetic mapping. Rather, genetic distance between two loci is a function of the density of occurrence of crossing-over in the interval they define. The relationship between recombination and genetic distance requires a mapping function which embodies assumptions about the crossing-over formation, as reviewed in detail in Bailey [1]. The lack of independence in crossing-over events, referred to as 'interference', is a major complication. Because the probability density of crossing-over along any chromosome is unknown, proper modeling of interference has been challenging. Indeed, most mapping functions proposed apply only to three-point tests.

Great attention was given to the formulation of models of interference from the 1930's to the 1960's, partly to solve the mapping problem, partly in hope that it could yield some insight into the nature of the biological process involved [1]. In recent years, the trend in human genetics has been to simply ignoreinterference altogether. A careful examination of the impact of this assumption on the inference of gene order [14] showed that, for the sample sizes that can be achieved in man, the simplification afforded by the assumption of independence in crossing-over far outweighed the moderate biases incurred in the process and the challenge of modeling adequately both interference and sex difference in recombination. More discussion of this matter can be found in White and Lalouel [36].

Under the assumption of no interference, a simple relationship holds between recombination and genetic distance, given by Haldane's mapping function [8]. Furthermore, multilocus-transmission probabilities can be expressed through simple recursive expressions. Because of incomplete data, it became apparent that linkage analysis in humans could derive more information by combining segregation information at multiple loci, leading to development of appropriate methods and computer programs [13, 15] which could be applied both to discrete and to continuous traits. Other algorithms and computer programs, tailored to particular modes of inheritance and simple family structures, were later introduced [4, 12].

Multilocus linkage analysis has been used to derive genetic maps of most human chromosomes. Determining the most likely gene order, and evaluating confidence in this order by assessing critical competing alternatives, present particular challenges because a combinatorial explosion occurs in the number of possible orders as the number of loci increases. A practical solution has required the implementation of algorithms for automated map construction [12, 16].

174

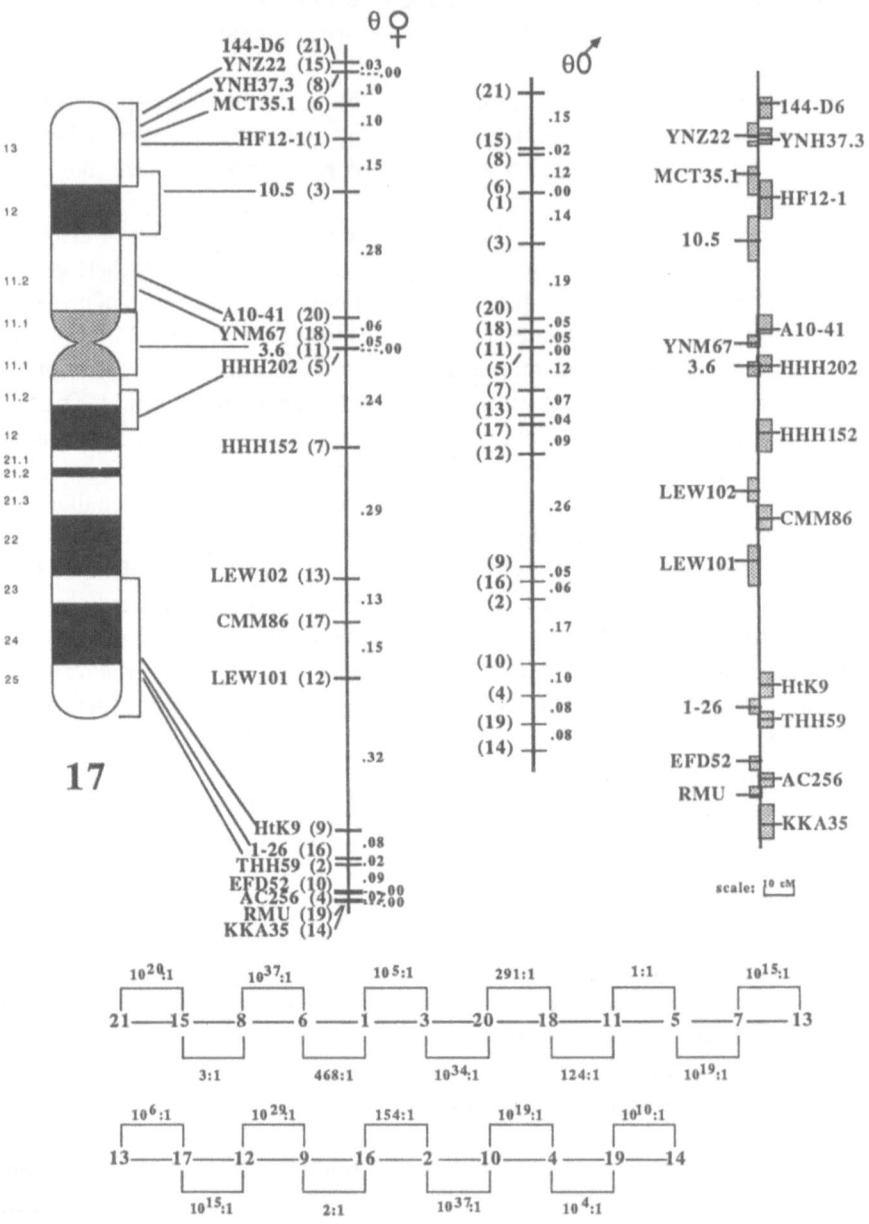

Fig. 1. Sex-specific genetic maps of chromosome 17, with physical localizations of selected markers indicated on the idiogram (left). Maps are scaled in centimorgans (cM) according to the Haldane mapping function, under the assumption of a variable sex ratio in each interval. θ = recombination fraction. Right: confidence limits (shaded bars) of maximum likelihood location for each of 21 loci on the map of chromosome 17. Genetic distances between loci can be derived from combined male and female recombination fractions. Below: odds against inversion of adjacent loci on the final map. (Reprinted from Nakamura *et al.* 1988, by permission of Academic Press.)

For illustration, a genetic map of chromosome 17 is presented in Fig. 1. This figure summarizes most of the relevant information required to evaluate the map and to use it for localizing an unknown gene or a new marker, including some measures of support for the reported order of loci. It should be emphasized that genetic distances obtained under the assumption of no interference do not provide good estimates of total map lengths. This is of little consequence, however, as what are likely to be of practical value are the recombination estimates read off the map for selected sets of loci.

The merit of multilocus linkage analysis also rests in the greater power it provides for detecting linkage between a genetic marker and a set of linked loci [12, 13, 32]. Once a map has been established, linkage between a new marker and this map can be sought by computing the likelihood ratio in favor of linkage for every location of the test locus along the map, generating a graph of support along the map, or location score [13]. This strategy efficiently uses all the information provided by the map and yields a visual representation of statistical support, as illustrated in Fig. 2. It can be applied to either a discrete or a continuous trait, as discussed earlier.

DNA markers for chromosome maps

Linkage analysis requires that genetic markers be available to detect linkage in any region of the genome. Initially, linkage studies in humans were performed with a battery of about 30 protein polymorphisms, most of which were of limited value because of their low heterozygosity. Despite this fact, numerous linkage studies were performed, and they yielded some clear successes.

The definition of polymorphism directly at the molecular level has removed this limitation. Since an initial proposal to systematically exploit the many DNA polymorphisms which can be revealed with restriction endonucleases [2], a large number of restriction fragment length polymorphisms (RFLPs) have been characterized in humans. Primary genetic maps have now been constructed for all human chromosomes, yielding quasi-complete coverage of the human genome at an average resolution of 10 to 20 centimorgans. These markers have made possible the assignment of a large number of inherited conditions of unknown etiology, such as Duchenne muscular dystrophy, Huntington disease, cystic fibrosis, adenomatous polyposis coli, and peripheral neurofibromatosis. Further cloning in restricted genomic regions defined by flanking DNA markers and/or by physical rearrangements has already led to the identification of the genes involved in the etiology of Duchenne muscular dystrophy [19] and cystic fibrosis [29, 30].

The DNA polymorphisms used in such studies arise from one of several mechanisms: (1) mutation which creates or destroys a restriction site recognized by an endonuclease; (2) insertion or deletion of a DNA segment; or (3) variation in the number of copies of tandem repeats of short DNA motifs (VNTRs). The last-named mechanism bears the greatest promise for linkage

176

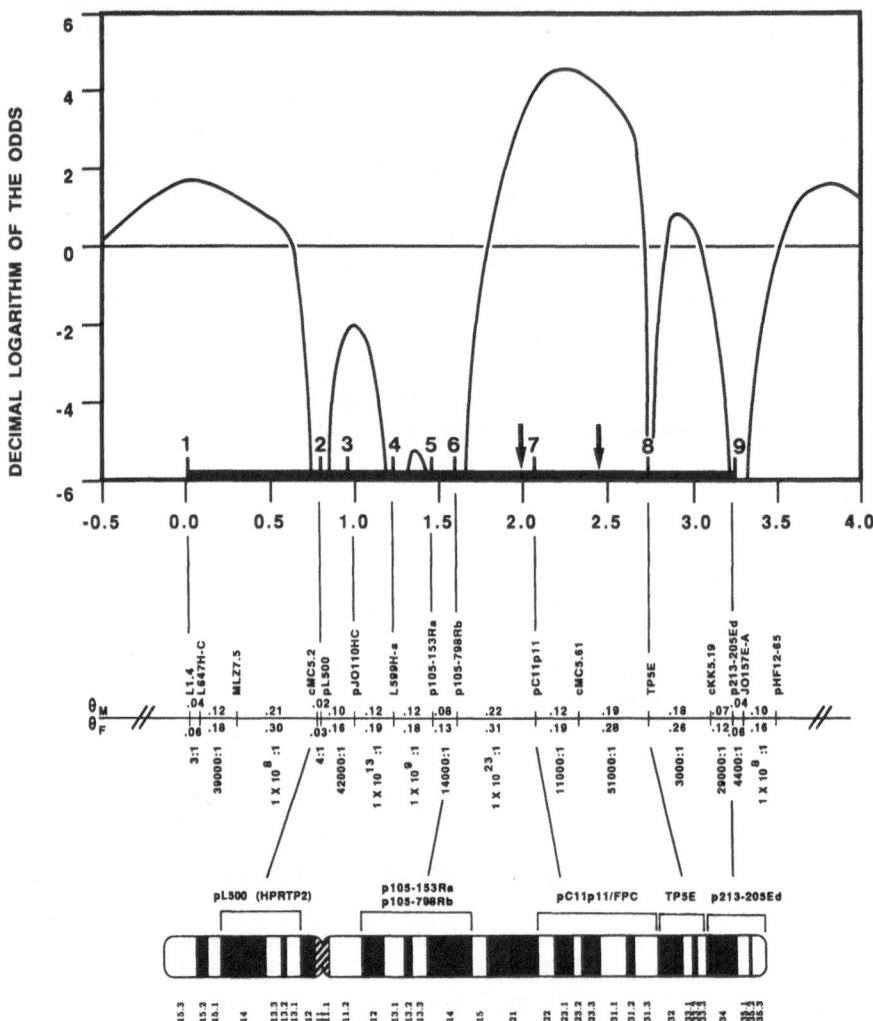

Fig. 2. Genetic linkage map of 16 DNA markers on chromosome 5, with location scores (top) for the locus harboring the mutation responsible for adenomatous polyposis coli (APC). Two arrows on the abscissa indicate the 95% confidence boundaries for the location of the APC gene on the map. The genetic map (center) indicates estimates of recombination distance (θ) between markers for males (above line) and females (below line). Physical locations of selected probes are indicated on the idiogram of the chromosome (below). (Reprinted from Leppert *et al.*, 1987, by permission of the American Association of the Advancement of Science.)

studies because it can generate a large number of alleles at a single locus, and heterozygosity within the population will tend to be high. Jeffreys [9] showed that a vast amount of polymorphism could be revealed at multiple loci by probes containing motifs of such repeats. In most experimental designs applicable to humans, such 'minisatellites' or 'genetic fingerprints' do not yield to easy interpretation, as allelic series cannot be readily defined. Nevertheless, screening of cosmid libraries with the so-called minisatellite probes has led to the isolation of a large number of clones that identify single-copy VNTR loci [24]. Furthermore, the probes that identify multiple loci may remain of merit for studies in plant or animal species where experimental control exists, because the numerous progeny necessary for interpretation of their patterns of inheritance can be obtained.

Other recent developments might greatly facilitate the construction of genetic maps in various species. Enzymatic amplification by the polymerase chain reaction [23] can be used for the rapid typing of polymorphisms occurring within known DNA sequences, without the need for Southern transfers; this method can be applied to investigate both site polymorphisms and VNTR loci [10]. The technique even allows the characterization of polymorphism at the level of a single cell [17], a development which may be relevant for mapping genomes. Short tandem repeats of simple motifs such as $(GT)_n$, which seem to occur at high density throughout the genome, have the potential of yielding a large number of allelic series with high heterozygosity [34]; such systems lend themselves to amplification by PCR when their flanking sequences are defined. Without doubt, very significant technological advances are under way, which should drastically reduce the challenge of generating genetic maps and exploiting them for the identification of loci that affect phenotypes of scientific or commercial interest.

179

References

1. Bailey NTJ (1961) Introduction to the Mathematical Theory of Genetic Linkage. Oxford: Clarendon Press
2. Botstein D, White R, Skolnick M, Davis R (1980) Construction of a genetic linkage map in man using restriction fragment length polymorphisms. Am J Hum Genet 32: 314-331.
3. Byers PH (1989) Disorders of collagen biosynthesis and structure. In: Scriver CR, Beaudet AL, Sly WS, Valle D (eds). The Metabolic Basis of Inherited Disease, Vol. 2, pp. 2805–2842. New York: McGraw-Hill.
4. Clayton J (1986) A multipoint linkage analysis program for X-linked disorders, with the example of the Duchenne muscular dystrophy and serum DNA probes. Hum Genet 73: 68–72.
5. Dausset J (1986) Le centre d'étude du polymorphisme humain. Presse Méd 15: 1801–1802.
6. Edwards JH (1960) The simulation of Mendelism. Acta Genet Stat Med 10: 63–70.
7. Elston RC, Stewart J (1971) A general model for the genetic analysis of pedigree data. Hum Hered 21: 523–542.
8. Haldane JBS (1919) The combination of linkage values, and the calculation of distance between the loci of linked factors. J Genet 8: 299–309.
9. Jeffreys A, Wilson V, Thein S (1985) Hypervariable 'minisatellite' regions in human DNA. Nature 314: 67–73.
10. Jeffreys AJ, Wilson V, Neumann R, Keyte J (1988) Amplification of human minisatellites by the polymerase chain reaction. Nucleic Acids Res 16: 10953–10971.
11. Lalouel J-M, le Mignon L, Simon M, Fauchet R, Bourel M, Rao DC, Morton NE (1985) Genetic analysis of idiopathic hemochromatosis using both qualitative (disease status) and quantitative (serum iron) information. Am J Hum Genet 37: 700–718.
12. Lander ES, Green P (1987) Construction of multilocus genetic linkage maps in humans. Proc Natl Acad Sci USA 84: 2363–2367.
13. Lathrop GM, Lalouel J-M, Julier C, Ott J (1984) Strategies for multilocus linkage analysis in humans. Proc Natl Acad Sci USA 81: 3443–3446.
14. Lathrop GM, Lalouel J-M, Julier C, Ott J (1985) Multilocus linkage analysis in humans: detection of linkage and estimation of recombination. Am J Hum Genet 37: 482–498.
15. Lathrop GM, Lalouel J-M, White R (1986) Construction of human linkage maps: likelihood calculations for multilocus linkage analysis. Genet Epidemiol 3: 39–52.
16. Lathrop M, Nakamura Y, Cartwright P, O'Connell P, Leppert M, Jones C, Tateishi H, Bragg T, Lalouel J-M, White R (1988) A primary genetic map of markers for human chromosome 10. Genomics 2: 157–164.
17. Li H, Gyllensten UB, Xiangfeng C, Saiki R, Erlich HA, Arnheim N (1988) Amplification and analysis of DNA sequences in single human sperm and diploid cells. Nature 335: 414–417.
18. MacLean CJ, Morton NE, Elston RC, Yee S (1976) Skewness in commingled distributions. Biometrics 32: 695–699.
19. Monaco A, Neve R, Colletti-Feener C, Bertelson C, Kurnit D, Kunkel L (1986) Isolation of candidate cDNAs for portions of the Duchenne muscular dystrophy gene. Nature 323: 646–650.
20. Morton NE (1955) Sequential tests for the detection of linkage. Am J Hum Genet 7: 277–318.
21. Morton NE (1956) The detection and estimation of linkage between the genes for elliptocytosis and the Rh blood types. Am J Hum Genet 8: 80–96.
22. Morton NE, Maclean CJ (1974) Analysis of family resemblance. III. Complex segregation of quantitative traits. Am J Hum Genet 26: 489–503.
23. Mullis KB, Faloona FA (1987) Specific synthesis of DNA in vitro via polymerase-catalyzed chain reaction. Meth Enzymol 155: 335–350.
24. Nakamura Y, Leppert M, O'Connell P, Wolff R, Holm T, Culver M, Martin C, Fujimoto E, Hoff M, Kumlin E, White R (1987) Variable number of tandem repeat (VNTR) markers for human gene mapping. Science 235: 1616–1622.
25. Nakamura Y, Lathrop M, O'Connell P, Leppert M, Barker D, Wright E, Skolnick M,

Kondoleon S, Litt M, Lalouel J-M, White R (1988) A mapped set of DNA markers for human chromosome 17. Genomics 2: 302–309.

26. Ott J (1974) Estimation of the recombination fraction in human pedigrees: efficient computation of the likelihood for human linkage studies. Am J Hum Genet 26: 588–597.

27. Ott J (1983) Linkage analysis and family classification under heterogeneity. Ann Hum Genet 47: 311–320.

28. Ott J (1985) Analysis of Human Genetic Linkage. Baltimore: Johns Hopkins University Press.

29. Riordan JR, Rommens JM, Kerem B-S, Alon N, Rozmahel R, Grzelczak Z, Zielenski J, Lok S, Plavsic N, Chou J-L, Drumm ML, Iannuzzi MC, Collins FS, Tsui L-C (1989) Identification of the cystic fibrosis gene: cloning and characterization of complimentary DNA. Science 245: 1066–1075.

30. Rommens JM, Iannuzzi MC, Kerem B-S, Drumm ML, Melmer G, Dean M, Rozmahel R, Cole JL, Kennedy D, Hidaka N, Zsiga M, Buchwald M, Riordan JR, Tsui L-C, Collins FS (1989) Identification of the cystic fibrosis gene: chromosome walking and jumping. Science 245: 1059–1065.

31. Smith CAB (1963) Testing for heterogeneity of recombination values in human genetics. Ann Hum Genet 27: 175–182.

32. Thompson EA (1984) Information gain in joint linkage analysis. IMA J Math Appl Med Biol 1: 31–49.

33. Wald A (1947) Sequential Analysis. New York: John Wiley.

34. Weber JL, May PE (1989) Abundant class of human DNA polymorphisms which can be typed using the polymerase chain reaction. Am J Hum Genet 44: 388–396.

35. White R, Leppert M, Bishop T, Barker D, Berkowitz J, Brown C, Callahan P, Holm T, Jerominski L (1985) Construction of linkage maps with DNA markers for human chromosomes. Nature 313: 101–105.

36. White R, Lalouel J-M (1987) Investigation of genetic linkage in human families. Adv Hum Genet 16: 121–228.

9. Statistical methodologies for mapping and analysis of quantitative trait loci

JOEL IRA WELLER

Institute of Animal Sciences, A. R. O., The Volcani Center, Bet Dagan 50250, Israel

Introduction

The idea of using genetic markers to locate the individual quantitative trait loci (QTL) responsible for variation in quantitative traits goes back nearly to the beginning of modern genetics [34]. However, until the last decade, application of this technique was limited by the lack of segregating markers in the populations of interest. The discovery of restriction fragment length polymorphisms (RFLPs), which occur in all natural populations, made this technique generally applicable for the first time [4, 42].

The objective of this chapter is to familiarize researchers using RFLP technology with the statistical methods that are useful for the mapping and analysis of quantitative trait loci with the aid of genetic markers. The first section will be a general description of the pertinent statistical methodologies available. Therefore readers with a more advanced knowledge of statistics may wish to skip this section. The second section explains how these methods are applied to the detection of QTL. The third section deals with methodologies for the exact mapping and utilization of QTL in breeding.

Description of statistical methodologies

In this section we will describe three methodologies that have been applied to the question at hand: least squares, method of moment estimation, and maximum likelihood. The general progression will be from the simpler to the more complex. It is not our goal to describe each method in detail, rather to explain the philosophy behind each method, the situations in which each method is applicable, and the advantages and limitations of each.

Least-squares linear model estimation

A linear model is one in which the dependent variable is a linear function of the independent variables, for example the effect of fertilizer and temperature on yield per hectare of tomatoes. The linear model can be expressed as follows:

181

J.S. Beckmann and T.C. Osborn (eds) Plant Genomes: Methods for Genetic and Physical Mapping, 181–207.
© 1992 *Kluwer Academic Publishers*.

182

$$Y = b_1(F) + b_2(T) + e \qquad (1)$$

where Y is yield per hectare, F is fertilizer given, T is temperature, b_1 and b_2 are the regression coefficients, and e is the residual of Y not explained by the effects included in the model. It is assumed that Y, F, and T have been measured on a number of plots, and the goal of the analysis is to estimate b_1 and b_2. The principle behind least-square-estimation is to find the parameter estimates that minimize the residual sum of squares. The residual sum of squares as a function of the other terms can be written as follows:

$$e^2 = [Y - b_1(F) - b_2(T)]^2 \qquad (2)$$

The least-square parameter estimates can then be derived by differentiating equation (2) with respect to b_1 and b_2, and equating these partial differentials to zero. The values for b_1 and b_2 that satisfy these equations are the least-square estimates of these parameters. The main objective of least-squares estimation is that the model should explain as much of the variation in the dependent variable as possible, thus a model with zero mean squared residuals perfectly 'fits' the data.

These estimates will have the additional property of unbiasedness. 'Unbiasedness' means that the expectation of the parameters being estimated should be equal to the actual parameter value. In other words, if the experiment was repeated a very large number of times, the mean of the estimates should tend towards the true values. Under the assumptions of normal and independent distribution of residuals, estimates that meet these conditions can be derived by the 'normal equations' [35]. Derivation and explanation of the normal equations requires a strong basis in matrix algebra, and will therefore not be covered in this chapter. Least-squares linear model analysis can be performed by PROC GLM of SAS [33].

The independent variables considered so far have been 'continuous', such as height, weight, temperature, etc. Dependent variables can also be discrete, such as genotype, block or strain. In the terminology of SAS [33] these are called 'class' variables. Assume that the experiment described above was performed in several different regions. If the effect of the different regions is ignored, it is possible that the regression estimates will be biased. We will therefore rewrite equation (1) to include a region effect:

$$Y = b_1(F) + b_2(T) + R_i + e \qquad (3)$$

where R_i is the effect of the ith region, and the other terms are as described above. Note that no regression coefficient is written in conjunction with R. This is because the region effects on yield have no relationship to the region numbers, which may be arbitrary with respect to Y. The normal equations can still be used to derive least-squares estimates for both the continuous and class effects.

Prior to the advent of powerful computers, the accepted methodology for analysis of class effects was analysis of variance (ANOVA), which was used to determine statistical significance. Clearly it is much more useful to derive least-squares estimates, rather than just concluding that a given effect is 'significant'. If the different effects included in the model are independently distributed, then the class effects can be estimated by the simple means of the different levels. In practice this is rarely the case, and the normal equations must be utilized.

Two additional types of effects that can be included in linear models are interactions and nested effects. Interactions can occur among continuous or class effects, or between the two types of effects. To illustrate the various alternatives, a second class effect, strain, is added to the model. Without interactions the model can be written as follows:

$$Y = b_1(F) + b_2(T) + R_i + S_j + e \tag{4}$$

where S_j is the effect of the jth strain, and the other terms are as defined previously. Assume that over all regions, strain 2 gives 2000 kg more fruit per hectare than strain 1. However, in region 1 this difference is greater than 2000 kg, and in region 3 the difference is less. Inclusion of an interaction effect between these two class effects will explain an additional fraction of the variance of the dependent variable, not explained by the main effects. Class by continuous interaction can be illustrated by the following example. Assume that over all regions, increasing fertilizer by 1 kg/ha increases yield by 1000 kg/ha, but in region 1 increasing fertilizer by 1 kg only increases yield by 300 kg. Inclusion of an F by R interaction will then increase the explanatory power of the model. This line of reasoning can be extended to cover continuous by continuous interactions.

Assume that within regions, plants are divided into blocks. Then it may be desirable to include the effect of block in the model. This is an example of a nested effect. The important difference between a hierarchical experiment design (nested) and the factorial design described above is that interactions cannot be estimated for the former.

Significance in linear models is computed by comparing various model sums of squares to the residual sum of squares. If the effects included in the model have no effect on the dependent variable, then the model sum of squares and the residual sum of squares, each divided by their respective degrees of freedom, should be equal, except for sampling variation. If the model explains some of the variation in the dependent variable, then the model sum of squares will be greater than the residual sum of squares, and the expected ratio of sums of squares will have an expectation greater than one. The deviation of this ratio from unity can be tested by the F-statistic. For a model with more than one effect, there are several methods to compute the appropriate sums of squares needed to test significance for the different effects. The specific methods and their applicability are beyond the scope of this chapter, which emphasizes estimation, rather than tests of significance (the interested reader is referred to [33], Chapter 9).

Limitations of linear model estimation

Least-squares estimates have the desirable properties of minimum residual variance and unbiasedness, and can be readily computed with commonly available software. Therefore this is the method of choice for estimation of effects, where applicable. Because of these desirable attributes, many researchers use linear model analysis even in situations in which this type of analysis is not appropriate. Not all problems can be phrased in terms of linear models. Furthermore, linear model analysis is based on assumptions, which do not hold under certain circumstances.

We will first explain the latter problem. Ordinary least-squared estimation assumes that the residuals are independently normally distributed with equal variance. It further assumes that the effects of the 'independent' variables are in fact independent from one another. Estimates derived from data that violate these assumptions will not have least-square properties. A few examples will be used to illustrate. Assume that an experiment with tomato plants is performed in two regions. If the variance between plots is greater in region 2, then the residuals will not have equal variance. By assuming equal variance, the residuals for the observations from plot 1 will be larger than they should be, while those from plot 2 will be smaller. Thus the estimates of effects will 'fit' the results from region 2 better than those of region 1. Another example of a situation that violates the assumptions of least-squares analysis is a dependent variable with a non-normal distribution. For example, in a cross between *L. esculentum* (domestic tomato) and *L. pimpinellifolium* (a wild progenitor), the distribution of fruit weight is highly asymmetric [50]. If this trait is analyzed by least-squared methodology, the parameter estimates will be based nearly completely on the high values, which also have high variance, while the effect of the majority of observations will be negligible. Distributions of this type result when the biological scale is different from the scale in which the trait is measured [15]. Often problems of scaling can be alleviated by transformation of the data. In this example the log of fruit weight will have a quasi-normal distribution.

We will now give two examples of non-linear models:

$$Y_1 = b_1(\log x_1) + e_1 \qquad (5)$$

$$Y_2 = b_2/x_2 + e_2 \qquad (6)$$

where Y_1 and Y_2 are each dependent variables, x_1 and x_2 are independent variables, b_1 and b_2 are regression coefficients, and e_1 and e_2 are the residuals not explained by the models. As in the previous examples, the objective is to derive least-square estimates for b_1 and b_2. As can be seen for these models, the dependent variables are no longer linear functions of the independent variables. Thus a least-squares solution cannot be derived by the normal equations. Because of the advantages of linear model analysis, a common approach is to rewrite the model in a linear form. This can be done by defining new variables, in the example above: $x_3 = \log x_1$, and $x_4 = 1/x_2$. Using these variables, equations (5) and (6) can be rewritten as linear models. The disadvantages are as follows: if the residuals had a normal distribution on the original scale, they will no longer have a normal distribution on the new scale, and the model will be in violation of the assumptions listed above. Furthermore, although the estimates for the newly defined variables may have least-squares properties, estimates for the original variables based on the transformed variables will not. Thus an estimate for x_2 computed as the inverse of the estimate for x_4 may be biased even though the estimate for x_4 is not.

To overcome these problems, estimation methods are available that do not depend on the linear model assumptions. The main alternative methods of estimation are non-linear least-square estimation, the method of moment estimation, and maximum likelihood estimation. We will deal briefly with the first two methods, and explain the third in more detail.

Non-linear models and moments method estimation

We explained in the previous section that least-square estimates are preferred because they have the desirable properties of minimum residual variance and unbiasedness. Thus *a priori* it would seem that non-linear least-square estimation should be the method of choice for analysis of non-linear models. In fact this is rarely the case. In the previous sections we explained that linear leastsquare estimates can be derived by the normal equations. However, there is no uniformly best method for obtaining least-square estimates for non-linear models. Furthermore, in non-linear models, the functions used to compute the parameter estimates are functions of the parameter values. Thus iterative solutions are necessary, in which the values for the parameters in the $i + 1^{th}$ iteration are functions of the estimated values of the ith iteration. Most methods of non-linear model estimation also require that the derivatives of the model be specified. Often this is not a trivial task. Finally there is no guarantee that the procedure will be able to successfully fit the model. These problems have severely limited the use of non-linear models to solve most practical statistical problems.

To explain the moments method we will first define the concepts of central and non-central moments. Assume a sample of n observations $(X_1, X_2, \ldots, X_n)$. The first central moment is defined as follows:

$$\Sigma(X_i)/n \tag{7}$$

where X_i represents the ith observation. This is of course the sample mean. The second central moment is defined as follows:

$$\Sigma(X_i - X.)^2/n \tag{8}$$

where X. is the sample mean. The second central moment is equal to the sample variance. In general, the mth order central moment is computed as:

$$\Sigma(X_i - X.)^m/n \tag{9}$$

The third central moment is an estimate of the skewness (asymmetry) of the distribution. The fourth central moment is an estimate of the kurtosis of the distribution, that is the degree to which the observations of the distribution are concentrated in the 'center' and the 'tails', as opposed to the 'shoulders' of the distribution. Non-central moments are computed in the same manner, except that some other value is substituted for the mean.

A class effect, such as strain, can be estimated by the first central moment. In plain language this means computing the mean production of each strain. A continuous effect can be estimated by the second central moment. Assume that the only effect on production is the quantity of fertilizer given. We can then describe the following equation:

$$\text{Var } Y = b_1^2 (\text{Var } F) \tag{10}$$

That is, the variance of yield per hectare is equal to the variance of fertilizer given times b_1^2. Thus b_1 can be estimated as $(\text{Var } Y/\text{Var } F)^{1/2}$. Although this estimate will not have the desirable properties of minimum residual variance and unbiasedness, there are some positive aspects of estimation by moments methods. First, computation is usually quite easy. It is not necessary to employ iterative techniques that are required for non-linear estimation. Also, the moments method is 'non-parametric'. That is, it is not dependent on assumptions about the nature of the distribution of the variables. Thus the moments method may be better than a parametric method for a situation where the assumptions of the parametric method are clearly violated.

Maximum likelihood estimation for a single parameter

The concept behind maximum likelihood (ML) is very simple and elegant, namely: find the parameter estimates that best match the sample of data. ML requires that the type of distribution from which the data were sampled be known. Thus ML, unlike the method of moments estimation, is a 'parametric' method of estimation. Unlike least squares, ML parameter estimates may be biased, and do not necessarily yield least-squares residuals. On the positive side, unlike the previous methods, ML is able to utilize all information available in the sample, it can be applied to both linear and non-linear models and, if performed correctly, always yields parameter estimates within the parameter space. This last property may seem trivial, but in fact for many complicated problems of estimation, other methods of estimation often yield parameter estimates outside the parameter space, for example, negative estimates of variance components.

The basic methodology for ML estimation will be illustrated using an example from a binomial distribution. Assume from a sample of ten observations, three are 'successes' and the other seven are 'failures'. We wish to derive the ML estimate (MLE) of p, the probability of success. This is done by writing the binomial probability of obtaining this result as a function of p:

$$L = \frac{10!}{3!\,7!}\,(p)^3(1-p)^7 \tag{11}$$

where L is the probability of obtaining this result, conditional on p. L is denoted the likelihood function. The MLE for p is that value of p which results in a maximum value for L. The MLE of p can be computed by differentiating L with respect to p, and solving for p with this derivative set equal to zero. In practice it is usually easier to compute and differentiate the log of L. This is equivalent to differentiating L, since a function and its log will have maximum value for the same variable value. It is thus possible to derive the MLE of p as follows:

$$\log L = \log(10!) - \log(3!\,7!) + 3(\log p) + 7[\log(1+p)] \tag{12}$$

$$d(\log L)/dp = 3/p - 7/(1-p) = 0 \tag{13}$$

$$p = 3/10 \tag{14}$$

This is of course the fraction of successes derived in the sample. Thus for this simple case, the MLE is the intuitive one.

From the above discussion, it should be clear why MLE must lie within the parameter space. A parameter estimate outside the parameter space will by definition have a likelihood of zero, and can therefore not be the MLE.

For a continuous distribution, the likelihood is computed as the statistical density of the distribution, conditional on the sample. Statistical density for a continuous variable is defined as the ordinate of the distribution function. For example, assume that a sample was taken from a normal distribution. To obtain the MLE for the mean, it is necessary to compute the joint statistical density of the sample. For a single observation the likelihood will be:

$$L = \frac{e^{-(X-\mu)^2/2\sigma^2}}{\sqrt{2\pi\sigma^2}} \tag{15}$$

where σ is the standard deviation, e is the base for natural logarithms and is approximately equal to 2.72, μ is the mean, and X is the variable value. For a sample of N observations, the likelihood will be the product of the likelihoods for each individual observation. As in the previous case, the MLE for μ can be derived by taking the derivative of the log of the likelihood, with respect to the mean, and setting this function equal to zero. The derivative of log L for a sample from a normal distribution is computed as follows:

$$d(\log L)/d\mu = \Sigma(X_i - \mu) \tag{16}$$

Setting this function equal to zero, we find that the MLE of the population mean is the sample mean, which is again the intuitively correct result.

The MLE for the variance could be derived in the same manner, and would again yield the intuitive result of the sample variance. Although in the two examples given so far, ML has been used to derive estimates that could have been derived by other methods, it will be demonstrated below that for more complicated problems, ML is the only method of estimation that can utilize all the available data.

The following equation can generally be used to derive the prediction error variance of the MLE:

$$\text{Var}\,(\hat{\phi} = \frac{-1}{E[d^2(\log L)/d\phi^2]} \tag{17}$$

where $\hat{\phi}$ is the MLE of ϕ, and $E[d^2(\log L)/d\phi^2]$ is the expectation of the second derivative of L with respect to ϕ. Equation (17) will be correct if the first derivative of ϕ is a multiple of the difference between the true parameter value and its estimate. Otherwise the prediction error variance will be slightly greater than the right-hand side of equation (17). Under a wide range of conditions, equation (17) will be asymptotically correct; that is, as the sample size increases the difference between the left-hand and right-hand sides of the equation tends towards zero. The square root of the prediction error variance will be the standard error of the estimate, and this value can be used to determine the confidence interval of the estimate.

Maximum likelihood multi-parameter estimation

ML can also be used to estimate several parameters simultaneously, for example, to estimate both the mean and variance in a normal distribution. In that case it is necessary to maximize the likelihood with respect to both parameters. This can be done by taking the partial derivatives of the log likelihood with respect to each parameter, and setting each derivative equal to zero. It is then necessary to solve a system of equations equal to the number of parameters being estimated. In general the likelihood function for estimation of m parameters, $(\phi_1, \phi_2, \ldots, \phi_m)$, from a sample of n observations $(X_1, X_2, \ldots, X_n)$ can be written as follows:

$$L = p(X_1, X_2, \ldots, X_n/\phi_1, \phi_2, \ldots, \phi_m) =$$
$$= p(X_1/\phi_1, \phi_2, \ldots, \phi_m)p(X_2/\phi_1, \phi_2, \ldots, \phi_m) \cdots p(X_n/\phi_1, \phi_2, \ldots, \phi_m)$$
$$= \Pi\, p(X_i/\phi_1, \phi_2, \ldots, \phi_m) \tag{18}$$

where $p(X, \phi)$ represents the probability of obtaining X, conditional on ϕ, and Π signifies the product $p(X_i/\phi_1, \phi_2, \ldots, \phi_m)$ from X_1 through X_n. If the distribution is continuous, then p(X, ϕ) will be replaced with f(X, ϕ), i. e., the density of X, conditional on ϕ. Thus any problem that can be phrased in terms of equation (18) can be solved by ML.

The prediction error variances for the multi-parameter estimation problem can be derived in a manner parallel to that described in equation (17). This will be illustrated with the aid of matrix algebra [36]. The parameter estimates and the first derivatives will each consist of a vector with the number of elements equal to m. The second derivatives and the prediction error variances will both be square m by m matrices. Using brackets to denote matrices and vectors, the matrix of prediction error variances can be computed with the following equations:

$$\text{Var}\,[\hat{\phi}] = -\left[\frac{d^2 \log L}{d[\phi]^2}\right]^{-1} \tag{19}$$

where the right-hand side is the inverse of the matrix of second partial derivatives with respect to $[\phi]$. The diagonal elements will be the prediction error variances of the estimates, and the off-diagonal elements will be the prediction error covariances between the elements. These are needed to test hypotheses based on linear functions of the parameters.

Even if the prediction error variance is not computed, ML can still be used to test hypothesis, by a 'likelihood ratio test'. In a likelihood ratio test the maximum likelihoods obtained under two alternative hypotheses are compared. Under the assumption of no difference, the natural log of the likelihood ratio will be asymptotically distributed as $(-1/2)X^2$, where X^2 is the chi-squared statistic with one degree of freedom [37].

Although it is generally possible to write the likelihood function, and differentiate log L with respect to the different parameters, it is often not possible to derive an analytical solution to the resultant system of equations. One solution to this problem is to use general-purpose algorithms that find the maximum of a function, but these tend to be inefficient for large samples, and do not work in all cases. Several methods have been developed specifically for ML. We will consider three methods that have been applied to problems of QTL mapping: Fisher's method of scoring [3], expectation maximization (EM) [12] and Newton-Raphson's method [10]. All of these methods are iterative. The parameter estimates of the ith iterate are computed by solving a system of equations equal in number to the number of parameters being estimated. These reduced equations are themselves functions of the parameter estimates from the previous iteration. Thus it is necessary to continue iteration until changes between rounds fall below a sufficiently small value. It should also be noted that all of these methods only find a local maximum. Thus if different starting values for the parameters are used, it is possible that convergence can be obtained at different parameter values.

In Newton-Raphson, both the first derivatives and the matrix of second derivatives must be computed. Solutions for the ith round of iteration are computed by solving the following system of equations:

$$[\hat{\phi}_i] = [\hat{\phi}_{i-1}] - \frac{d(\log L)}{d[\phi]} \left[\frac{d^2 \log L}{d[\phi^2]} \right]^{-1} \qquad (20)$$

where $[\hat{\phi}_i]$ is the estimate of ϕ for the ith iterate, $[\hat{\phi}_{i-1}]$ is the previous estimate of $[\phi]$, and the other terms are as defined above, with derivatives computed for the $i - 1$ estimate of $[\phi]$.

The main advantage of Newton-Raphson is that convergence is generally rapid. The disadvantages are that the algorithm may not converge, even if the likelihood does have a maximum within the parameter space, and computation of the matrix of second derivatives is often not a trivial task. However, as shown above, this matrix can be used to derive estimates of the standard errors of the estimates, which is itself an important objective.

EM is generally considered the method of choice because it is guaranteed to converge to a maximum, provided that one exists within the parameter space. The rate of convergence, however, may be very slow. The principle behind EM is to consider two sampling densities, one based on the complete-data specification (unknown), and the second based on the incomplete-data specification (known). For example, assume that the data are derived from a mixture of two normal distributions with equal variances, but unequal means. The objective is to estimate the means of the two normal distributions, their variance, and the proportion of observations sampled from each distribution. The incomplete-data density function can be written as: $f(X_1, X_2, \ldots, X_n/\phi_1, \phi_2, \ldots, \phi_m)$. This density function is 'incomplete' in the sense that, for each observation, the normal distribution from which it was derived is unknown. The complete-data density function can be written as follows: $g(Y_1, Y_2, \ldots, Y_n/\phi_1, \phi_2, \ldots, \phi_m)$, where $Y_1, Y_2, \ldots, Y_n$ are observed only indirectly through $X_1, X_2, \ldots, X_n$. In this case $Y_1, Y_2, \ldots, Y_n$ is the vector of observations, with the additional information that the origin of each observation with respect to the two normal distributions is known. The objective of the EM algorithm is to maximize $f(X_1, X_2, \ldots, X_n/\phi_1, \phi_2, \ldots, \phi_m)$ with respect to $\phi_1, \phi_2, \ldots, \phi_m$, using the function $g(Y_1, Y_2, \ldots, Y_n/\phi_1, \phi_2, \ldots, \phi_m)$.

The EM algorithm consists of two steps: the estimation step, in which the 'sufficient statistics' are estimated for the complete-data density function; and the maximization step, in which this function is maximized with respect to the parameters. The 'sufficient statistics' are those statistics that must be computed from the data in order to maximize the likelihood. In the case of a mixture of normal distributions, the 'estimation step' consists of estimating the proportion of observations sampled from each distribution, based on the parameter estimates from the previous iteration. Using this estimate, the 'complete' likelihood function can be written. The maximization step then consists of computing the ML parameter estimates for the 'complete' likelihood function with respect to all parameters. Thus, it is still necessary to solve a system of equations equal to the number of parameter being estimated. The advantage of this procedure is that for many important theoretical distributions, the system of equations derived by equating the derivatives of the expectation of the complete-data sufficient statistics to zero is amenable to solution.

The specific equations necessary to apply EM, and the proof of convergence within the parameter space, are rather complicated, and will therefore not be presented in this chapter (the interested reader is referred to [12] for a complete discussion of the EM algorithm).

Detection of segregating quantitative trait loci

A large number of experimental designs and statistical methodologies have been suggested to detect the individual genes affecting quantitative traits with the aid of genetic markers. All of the experimental designs postulated have several elements in common, and these will now be reviewed briefly. Assume that the putative QTL is genetically linked to a marker locus, and that a particular individual is heterozygous for both loci. The genotype of this individual for the marker loci will be denoted M1M2, and for the QTL A1A2. Half of this individual's progeny will receive the allele M1, and half M2. If the allele A1 is located on the same chromosome as M1 then, except for recombinants, those progeny that receive M1 will also receive A1; while those individuals that receive M2 will also receive A2. Thus the effect of the QTL can be detected by comparing the mean of the two progeny groups that receive the alternative alleles from the heterozygous parent. The various designs suggested differ in the methods used to create the heterozygous parent, and the crosses performed.

The statistical methods used to detect QTL have generally been parametric. That is, they were based on assumptions as to the nature of the distributions of the observations. We will therefore first describe the 'usual' assumptions used to detect QTL, the types of effects postulated, and the types of data sets used. We will then describe the various experimental designs that have been used or postulated to detect QTL via linkage to genetic markers.

Assumptions, problems and types of effects postulated

Most studies that have dealt with the detection of linkage between QTL and genetic markers have not been careful to rigorously list the assumptions on which they based their analyses, although there are exceptions [49, 50, 54, 55]. A number of assumptions have been employed in most analyses, and we will list these 'usual' assumptions. We will also note the studies that attempted to remove or test some of these assumptions.

First, nearly all analyses have assumed that, except for the effect of the segregating QTL, the underlying distribution of the trait was normal. This assumption is required for both ANOVA and ML analyses. The only analyses that did not require this assumption are [54, 55], which used the method of moment estimation, and [28], which used chi-squared analysis. As explained above, these methods are non-parametric, and do not depend on the nature of the distribution. Theoretically, ML could be employed if some alternative distribution was postulated, but this has not been done in practice.

Methods to test for deviation from normality are available, but are not powerful for samples of moderate size. One study tested 18 quantitative traits, and found a number of traits with significant skewness and kurtosis, and one trait had significant kurtosis even though the distribution was symmetric [50, 53]. As described above, measuring a trait on some scale other than the biologically significant one can result in a skewed distribution. This problem can generally be alleviated by either a power or logarithmic transformation of the data. Both types of transformations were employed to obtain distributions with virtually zero skewness [50].

Other assumptions generally employed deal with the generally accepted principles of Mendelian genetics, i. e., equal probability that either chromosome of a pair will be transmitted to the zygote, and random association among homologous chromosomes in meiosis. The generality of these principles is supported by a large body of data, although exceptions have been found in certain cases, for example meiotic drive [32]. In addition, equal viability of all genotypes, and complete penetrance for the genetic marker have been generally assumed. In one test of these assumptions, five out of ten genetic markers displayed significance deviations from the expected Mendelian ratios [53].

Most studies that have attempted to map QTL have in addition assumed that there was only a single QTL linked to the genetic marker, that the genetic marker does not have a pleiotropic effect on the quantitative trait, and that only two QTL alleles were segregating in the test population. The final assumption is probably not problematic for crosses between highly inbred lines, since each line should be homozygous for a single allele. It may however be a significant problem for analysis of populations which are either outbred, or only moderately inbred.

The case of two loosely linked QTL was studied [27], although the analysis was based on a model constructed on the assumption of a single segregating QTL. Weller, Soller, and Brody [53] employed the heuristic argument that if the number of detectable segregating QTL is low, and if these loci are randomly distributed, then the probability of two loci being genetically linked will be low. It should also be noted that two tightly linked QTL can be considered as a single locus for most experimental designs employed.

Most analyses have assumed that the effect of the QTL on the quantitative trait was additive. That is, the only effect of allele substitution is on the mean of the quantitative trait. A few studies have considered QTL variance effects, and a rather large number of significant effects have been found [14, 53, 55]. An effect on the trait variance can be considered a multiplicative effect. No studies have considered higher-order QTL effects. For example, if the effect of gene substitution is non-linear, then the QTL could effect either the skewness or kurtosis of the distribution.

Finally, most analyses have ignored both within-loci interactions (dominance), and between-loci interactions (epistasis), even though significant dominance and epistasis effects have been found [14, 24, 44, 45, 53]. It should also be noted that dominance effects in an F-2 design will also effect the within-marker-genotype variance for the quantitative trait [2, 49].

Experimental designs for detection of QTL in crosses between inbred lines

Most analyses performed to detect QTL have been based on planned crosses, although some studies have been carried out on existing animal populations, especially dairy cattle [1, 6, 8, 19–22]. We will first discuss the simpler case of crosses between inbred or haploid lines. As stated above, the first step is to cross two lines differing in genetic makers to produce heterozygous F-1 progeny. After this step the following progeny types have been considered for analysis [25, 37, 41, 43, 47–50, 53–55]:

1. F-2 individuals produced by selfing the F-1 individuals, or crossing among them.
2. Backcrossing (BC) the F-1 individuals to one of the parental strains.
3. Recombinant inbred lines produced from single F-2 (RIF) individuals, or brother-sister matings.
4. Recombinant inbred lines produced from single backcross (RIB) individuals, or brother-sister matings.
5. Doubled haploid (DH) lines produced by self-fertilizing doubled haploids derived from the F-1.
6. Testcross (TC) progeny produced by mating the F-1 individuals to a third inbred line.

More complicated designs can be considered, but these are basically variations of the designs listed above. Biological, economic, and statistical considerations have been used to choose the experimental design. Biological considerations are that not all designs are possible with all species, for example DH. For certain species with almost complete self-fertilization, it is much easier to produce large numbers of F-2 individuals than BC or TC, which require cross-fertilization. On the other hand, for outbreeding species, inbreeding can result in reduced fitness, due to the presence of recessive deleterious genes in the population. Economic considerations refer to the economic value of the possible crosses. For example, if the goal is to introgress genes for a specific trait from a wild strain into a cultivar, the BC to the cultivar will probably have much greater economic interest than the F 2. This is also the major reason for the TC-type analyses. It should be noted, however, that dominance relationships cannot be estimated from BC or TC analyses. Statistical considerations refer to selecting a design with the maximum power to detect QTL effects. For example, in the RI and DH experimental designs, all individuals within the line will have the same genotype. Thus it will be necessary to genotype only a single individual of each line. However, the phenotypic performance of all individuals can be used to determine the QTL effect. This reduces the residual variance by a factor of n where n is the number of individuals measured in each line, and assuming that the other sources of variance are small, increases statistical power by a factor of $\sqrt{n}$. Since only the residual variance is decreased by either of these designs, the advantage of these techniques decreases as a function of heritability and recombination frequency between the QTL and the genetic marker.

Linear model analysis of crosses between inbred lines

Most statistical analyses have been by a variation of the following linear model:

$$Y_{ijk} = G_i + B_j + e_{ijk} \tag{21}$$

where Y_{ijk} is the trait value for the kth individual of the jth 'block', and the ith genotype; G_i is the effect of the ith marker genotype, B_j is the effect of the jth 'block' and e_{ijk} is the random residual associated with each individual. The 'block' effect represents all environmental effects that groups of individuals may have in common, such as row, field block, and season of growth. A significant genotype effect is indicative of a segregating QTL affecting the quantitative trait.

The assumptions and advantages of least-squared analysis have been described above, the main consideration being that this type of analysis can be readily performed by most commonly used statistical packages. The disadvantages are as follows:
1. The analysis does not estimate the recombination distance between the marker locus and the QTL. Thus the estimated effect is confounded with the effect of recombination and is a biased estimate of the actual QTL effect.
2. The method does not distinguish between a linked QTL, and a pleiotropic effect of the genetic marker.
3. The statistical analysis does not account for a situation of multiple comparisons, i. e. the effect of several genetic markers on several quantitative traits.

The first two objections will be considered in the final section of this chapter. The problem of multiple comparisons can be explained as follows. Assume that ten independent marker-trait combinations are tested by the model of equation (21). If a significance level of 0.05 is required for each comparison, then the probability that at least one of the comparisons will display a 'significant' difference purely by chance is: $1 - 0.95^{10} = 0.4$, which is much higher than 0.05. Several studies have addressed the problem of multiple comparisons, although no commonly accepted solution has been presented [27, 53]. If the comparisons are in fact independent, then to obtain a 'true' significance level would require a nominal significance level that satisfies the following equation:

$$p = 1 - (1 - p')^n \tag{22}$$

where p is the 'true' significance level, p′ is the nominal significance level, and n is the number of comparisons. In the example given above, to obtain a 'true' significance level of 0.05 with ten comparisons, requires a nominal significance level of 0.005. However requiring a higher nominal significance level would mean that some real effects will be considered as non-significant. A possible solution would be a joint analysis of all marker-trait combinations, in which each segregating marker is considered an effect, and each quantitative trait is considered a dependent variable in a multivariate analysis. This type of analysis can be performed by available statistical packages, such as SAS, but has not been reported in the literature. The disadvantage of this analysis is that inclusion of a large number of effects will decrease the number of degrees of freedom. Thus it will be necessary to sample a large number of individuals so that there will be sufficient degrees of freedom to perform the multimarker analysis.

The problem is further complicated by the fact that in practice neither the quantitative traits nor the genetic markers are independently distributed. The problem of correlations among the quantitative traits can be handled in theory by a multivariate analysis, provided that the residual covariance among the traits is known, or can be estimated. Lander and Botstein [27] considered both the 'sparse map' and 'dense map' situations in which either a few or a large number of segregating genetic markers are scattered over the genome. They derived an expression to compute the 'nominal' significance level required as a function of the 'true' significance level, the number of chromosomes, and the total length of the genetic map of the species.

Estimation of QTL effects in outcrossing populations

For most species of fruit trees or domestic animals it is impractical to produce the inbred lines which are the basis of the experimental designs described above. Instead, experimental designs have been based on the analysis of many progeny of a single or a few heterozygous individuals [5]. This type of analysis has been used chiefly for dairy cattle in which a single sire can have hundreds or thousands of progeny with records on a number of quantitative traits, while each dam will have only a few progeny. If only a progeny of a single parent is considered, then the experimental design is parallel to the testcross design described above, the only differences being that the analysis is based on a single F-1 individual (the heterozygous parent) and that this individual is mated to a random sample from the population, instead of to a third inbred line. A marker-linked segregating QTL can be detected by analysis of the progeny records by the linear model given in equation (21), the only difference being that G_i now represents that allele that each progeny received from the heterozygous parent.

There are two main problems with this type of analysis. First, if either of the alleles of the heterozygous parent are common to the population, a large fraction of the progeny will have the same heterozygous genotype as the parent.

Since it will not be possible to determine which allele was passed by the common parent, records of these progeny have generally been deleted. However, if a QTL is segregating in the population, it will also be segregating in the individuals that are heterozygous for the genetic marker, and will therefore increase the variance of the heterozygous genotype, as compared to the two marker homozygotes. Recently Dentine and Cowan [13] have described a method based on ML to include information from these heterozygotes in the analysis.

Second, even if a QTL is segregating in the population, the specific individual analyzed may be homozygous for the QTL. To overcome the second problem, most studies have been based on analysis of number of heterozygous parents. In this case analysis by the model of equation (21) can lead to incorrect conclusions. Even if several of the individuals analyzed are heterozygous for a marker-linked QTL, the linkage relationships may be different for different individuals. Thus, summed over all progeny groups, there will be no effect associated with the segregating marker alleles. This problem can be solved by analysis with the following model [40]:

$$Y_{ijkl} = S_i + G_{ij} + B_k + e_{ijkl} \tag{23}$$

where S_i is the effect of the ith parent, G_{ij} is the effect of the jth allele, nested within the ith parent, and the other terms are as defined above. Although this analysis solves the problem of differing linkage relationships, it has the disadvantage that a different effect is estimated for each marker allele within each sire family. If in fact there are only two QTL alleles segregating, then G_{ij} should assume only one of two values. However, due to sampling error, the estimates for G_{ij} will cover a broad range, and may not even display bimodality. It may be possible to determine the QTL genotype of individual parents by application of a multiple-range test, such as the Duncan test [33], for the G_{ij} effect. The power of this method to distinguish between heterozygous and homozygous individuals for the QTL has not been estimated.

Alternatively, significance can be estimated by computing the mean within-parent deviation between the two progeny groups with opposing marker alleles, and dividing these differences by their standard errors. The sum of squares of these deviations will have a chi-squared distribution, with degrees of freedom equal to the number of parents [18, 30]. Power was estimated for both ANOVA [41] and chi-squared analysis [51].

Alternatively, RFLP genotypes can be determined on a sample of progeny of a heterozygous parent, and the quantitative traits can be scored on the grandprogeny (i. e. the progeny of the RFLP-genotyped offspring of the heterozygous parent) [51]. This design is advantageous if the cost of growing and scoring individuals for the quantitative traits is significantly less than the cost of RFLP genotyping. This design is termed a 'granddaughter' design, as opposed to the 'daughter' design described. A segregating marker-linked QTL can be detected with analysis by the following linear model:

$$Y_{ijklm} = GS_i + G_{ij} + SO_{ijk} + B_l + e_{ijklm} \qquad (24)$$

where GS_i is the effect of the ith grandparent, SO_{ijk} is the effect of the kth individual with the jth marker allele, progeny of the ith grandparent, and the other terms are as defined for equation (23). As in the previous design, a significant marker-allele effect will be indicative of a linked QTL. Increasing the number of grandprogeny will reduce the residual variation, but not between-progeny genetic variation. Thus the advantage of the granddaughter design is greater for low-heritability traits.

Finally, Fernando and Grossman [16] developed a method to estimate QTL effects in any population structure, provided that the heritability and recombination frequency between the QTL and the genetic marker are known. Their method is an extension of Best Linear Unbiased Prediction (BLUP) via the 'animal model'. However, in addition to the restrictions listed above, this method requires solving a system of equations of rank greater than three times the number of individuals scored in the analysis. Cantet and Smith [9] demonstrated that the number of equations can be significantly reduced by absorbing the equations for non-parent individuals. For plant species with many progeny per mating, this could be a significant saving in computing requirements.

The statistical power of alternative designs

Statistical power to detect segregating QTL will depend on the sample size, the magnitude of the Type I error allowed, the effect of the segregating QTL in comparison to the genetic and environmental variances, the recombination distance between the QTL and the genetic marker, the specific experimental design employed, and the method of statistical analysis. Since the number of possible combinations described is quite large, we will present only a few examples from the literature, and describe in general terms the effect of various parameters on the statistical power of the experiment.

The magnitude of the QTL effect that can be detected will of course be a function of the residual variance of the model. However, since residual variance is only known *a posteriori*, we will in the remainder of this chapter consider QTL effects in units of the phenotypic standard deviation (SDU). Statistical power of 0.9 to detect a tightly marker-linked QTL that accounts for 1% of the phenotypic variance in the F-2 design, with a type I error of 0.05, is obtained with a sample size of about 1000 individuals [41]. The mean difference in trait values between alternative homozygotes for a QTL of this magnitude will be 0.282 SDU. For a power of 0.5, and 0.1 recombination, about 500 F-2 recombinant inbred lines are required to detect a QTL of similar magnitude, if only one individual is scored per inbred line [37]. For the daughter design a power of 0.7, with a type I error of 0.01, is obtained for a QTL of 0.2 SDU if 400 progeny of each of parent are analyzed for a trait with a heritability of 0.2; while for the granddaughter design, power is 0.95 if genetic markers are analyzed on

100 progeny of each of 20 grandparents, with 50 quantitative trait-recorded grandprogeny/progeny [51].

Statistical methods that utilize more information in the data, and make more extensive assumptions will be more powerful than methods that utilize less information and make less assumptions. Thus parametric tests, such as ANOVA or ML, will be more powerful than nonparametric methods, such as chi-squared analysis or MME; and ML, which utilizes all information in the data, will be more powerful than ANOVA, which utilizes only the mean and variance of the distributions.

The effects of the magnitude of the QTL and the proportion of recombination between the marker and the QTL on sample size to achieve a given power will be quadratic. That is, for an effect of half the magnitude, it will be necessary to increase the number of individuals scored four-fold to achieve the same power. In either the F-2 or BC designs, the magnitude of the effect measured will decrease proportional to $1-2r$, as compared to complete linkage, where r is the recombination proportion. Thus to achieve power equal to the case of complete linkage, it will be necessary to increase the experiment size by a factor of $1/(1-2r)^2$ [27, 41]. For QTLs bracketed by two markers, the effect measured will not be reduced by recombination (except for double cross-overs), but, in a simple linear model analysis, recombinant individuals will be deleted. The proportion of recombinants for the F-2 and BC designs will be $(1 - r_1)^2$ and $(1 - r_1)$, respectively, where r_1 is the recombination frequency between the markers. Power with a marker bracket will therefore be reduced by this factor relative to complete linkage. Assuming $r_1 = 2r$, which is the optimum case for a marker bracket, power with a marker bracket will be increased by $(1 - r_1)$ for the BC design, and will be equal to a single-marker analysis for the F-2 design [M. Soller, personal communication]. (In more complicated analyses, it will be seen that additional information can be derived from the recombinants.)

The power of those designs which are based on measuring the quantitative traits on several individuals derived from a single individual scored for marker genotype, such as RIF or the granddaughter design, will vary inversely with heritability. Using DH or recombinant inbred lines with minimal recombination, power equal to an F-2 design is obtained with only a third of the number of RFLP analyses for traits of moderate heritability ($h^2 = 0.2-0.3$) [43]. However this advantage decreases to only 2-fold if either heritability is greater than 0.5, or recombination frequency is above 0.2. In conclusion, the preponderance of literature sources indicate that for most of the designs considered, an experiment consisting of several hundred to several thousand individuals will be necessary to detect a QTL with an effect of 0.1 to 0.5 SDU.

Mapping and utilization of quantitative trait loci

Until now we have only considered the problem of detection of segregating QTL. Once economically useful loci have been detected, the goal will be to incorporate the desirable alleles into domestic cultivars. This will generally require that the position of the QTL be known more precisely. Therefore, this final section is divided into statistical methods for precise mapping of QTL, and methodologies for genetic manipulation of QTL.

Mapping QTL by linkage to a single marker

Two methods have been described in the literature for mapping QTL linked to a single marker, MME [54, 55] and ML [49]. MME has the advantages described above, i.e., it is not dependent on the distribution of the quantitative trait, and is relatively easy to compute, but it does not efficiently utilize all the information available in the sample. The principle behind mapping a QTL by MME is that incomplete linkage will result in a skewed distribution of the individual marker-genotype samples. Since skewness is measured by the third moment, a difference in skewness can be used to estimate linkage distance between the QTL and the genetic marker.

This method will be illustrated using the example of a BC design. Two marker-locus genotypes will be generated in the BC, M1M1 and M1M2. If linkage between the QTL and the marker is complete, then all M1M1 individuals will also be A1A1, and all M1M2 individuals will also be A1A2. If linkage is incomplete, then a proportion, r, of the M1M1 individuals will have genotype A1A2 for the QTL, while the same proportion of M1M2 individuals will have the QTL genotype A1A1. Thus, the marker-genotype distributions will be a mixture of two distributions for the quantitative trait, and if r is less than 0.5, these distributions will be skewed in opposite direction. The degree of skewness will be a function of r. Thus, it is possible to construct a series of equations in which the basic statistics of the marker-genotype distributions (means, variances, and skewness) are expressed in terms of the means and variances of the QTL genotype distributions and r.

In practice, this method has not tended to yield accurate estimates, and in fact produced many estimates outside the parameter space, for example negative estimates for r or the variances [54, 55]. Because of these problems, ML, which by definition can only yield estimates in the parameter space, has become the method of choice for QTL mapping.

We will illustrate ML also for the BC design, under the assumption of equal variances for the QTL. The statistical density function for a single individual of genotype M1M1 will be:

$$f(X) = \frac{(1-r)e^{-(X-\mu_1)^2/2\sigma^2}}{\sqrt{2\pi\sigma^2}} + \frac{(r)e^{-(X-\mu_2)^2/2\sigma^2}}{\sqrt{2\pi\sigma^2}} \qquad (25)$$

where X is the trait value, σ is the standard deviation, μ_1 is the mean of individuals with the A1A1 genotype, and μ_2 is the mean of individuals with the A1A2 genotype. Individuals with the M1M2 genotype will have the same likelihood, except that the means values will be reversed. The complete likelihood for a sample of individuals can be written as follows:

$$L = \prod^{n_1} [f(X_i, M1)] \prod^{n_2} [f(X_j, M2)] \tag{26}$$

where $\prod$ represents the product of a series, $f(X_i, M1)$ and $f(X_j, M2)$ are the statistical densities for i^{th} and j^{th} observations with genotypes M1M1 and M1M2, respectively; and n_1 and n_2 are the number of individuals with the two genotypes, respectively.

To obtain the ML parameter estimates, the log of this function must be differentiated with respect to four parameters, μ_1, μ_2, σ, and r. The partial derivatives must then be equated to zero, and this system of four equations must be solved. However, this system of equations can not be solved analytically. Since the number of parameters is relatively low, it should be possible to scan the four-dimensional space to find an optimum. This solution is not practical for the F-2 design, in which there will be an additional parameter, or for either design if the model assumes unequal variances. It should be noted, though, that if unequal variances are assumed, then the likelihood will tend toward infinity when either of the variances tends toward zero [29]. In practice, however, this has not been a problem, and a local maximum within the parameter space has been found.

Weller [49] suggested a combination of MME and ML for the F-2 design in which the means and variances for the homozygotes are estimated by MME and the likelihood is scanned only for the remaining three parameters: r and the mean and variance of the heterozygote. Simpson [37] used a general maximization algorithm to find a maximum for an the BC recombinant inbred lines design.

Mapping of QTL with marker brackets

If the putative QTL is bracketed by two segregating genetic markers, then both least squares and ML can be used to map the QTL. Various studies have shown that, if the frequency of double-crossing-over is assumed to be negligible, unbiased estimates for the QTL effect can be obtained by a linear model analysis of the non-recombinant types [48]. The ratio between the effects measured for the recombinant, and non-recombinant types can then be used to estimate the recombination frequency [50].

Lander and Botstein [27] used EM to find MLE for the BC design with marker brackets. Instead of maximizing jointly for all parameters, they found ML values of the means and variance for a given value of r, and then scanned the likelihood for the r value with the highest likelihood. A complete EM solution to this problem has not been derived, and it is possible that this problem is not amenable to solution by this algorithm.

Knapp *et al.* [25] demonstrated that with marker brackets, a constrained linear model can be used to estimate the recombination frequency for a number of different experimental designs. Non-linear models can be used to obtain simultaneous estimates of both r and the QTL means. These estimates can be derived with PROC-NONLIN of SAS [33], and will be the MLE, under the assumption of equal QTL variance.

Jenson [23] used Fisher's method of scoring [3] based on partial first and second differentials estimated by simple numerical methods to estimate the parameters of a DH design. Darvasi [11] used Newton-Raphson iteration to solve for the same case described by Lander and Botstein [27]. Maximization was jointly found for all parameters, and convergence was obtained with under ten rounds of iteration for all simulations. The main obstacle for wider application of this technique is that it requires calculation of both the vector of first differentials, and the matrix of second differentials, which is not a trivial undertaking, especially for more complicated designs.

All of these studies also considered either the confidence limits of the estimates or tests of significance for r against the null hypothesis of complete linkage [49, 50] or random association [37]. Weller [49], and Lander and Botstein [27] suggested scanning the likelihood function relative to r to determine a quasi-confidence interval for r. Simpson [37] tested the MLE for r against the hypothesis of random association, by a likelihood ratio test. Jenson [23] and Darvasi [11] used the matrix of second differentials to estimate standard errors, as described above.

Within- and between-strain genetic manipulation of quantitative trait loci

Marker-assisted selection (MAS) within a strain can be used in three ways: 1) to increase the accuracy of evaluation, 2) to increase selection intensity, and 3) to decrease generation interval. Smith and Webb [39] and Smith and Simpson [38] estimated the expected genetic advancement by selection on QTL due to more accurate genetic evaluation. They found that gains were minimal for high-heritability traits, but significant for low-heritability traits.

In trait-based selection of progeny of a cross between two lines, desirable alleles might be lost by chance. If the specific QTL accounting for most of the variance between two strains were known *a priori*, strategies could be employed for the most efficient incorporation of desirable alleles into a group of selected individuals. Weller and Soller [52] studied strategies for incorporation of a number of favorable genes into a single strain. They found the optimum strategy to be random mating among the different strains and mass selection on the number of positive alleles in each progeny. By this method, the most rapid gain is made in the first few generations. However, selection on marker genotype should be only slightly more effective than trait-based selection [26, 38]. It should be noted, though, that with trait-based selection, some of the favorable alleles will be lost after several generations. Using MAS it is possible to produce individuals homozygous for the favorable allele for a relatively large number of loci in eight to ten generations.

Although dominance, heterotic and epistatic effects have been found for individual QTL [14, 24, 44, 45, 53], very little has been published as to the utilization of this source of variation in breeding programs. Genotyping of QTL-linked genetic markers may allow for the utilization of nonadditive sources of genetic variance, especially in between-breed crosses.

Selection in domestic species is nearly always based on an index of several traits. Falconer [15] noted that during the course of a breeding program, negative genetic correlations will develop among the traits under selection. Those alleles with positive effects for both traits will be the first to undergo fixation, while those alleles with positive effects on some traits, and negative effects on others, will remain at intermediate frequency. Genetic progress for an index consisting of traits with major negative genetic correlations will be slow even if all traits have high heritability. Since *a priori*, there is no reason to assume that the negative genetic correlation will be distributed evenly over all QTL affecting the traits in question, it should be possible to locate specific loci with positive effects on both traits. Selection for these loci may be significantly more effective than traditional selection index. Previous studies have found QTL with effects on pairs of traits in opposite direction to the trait-based genetic correlation [53]. At present, no study has estimated the gain expected by marker-assisted selection in the case of index selection.

Often a breeder wants to transfer desirable alleles from one strain to another. This process, called 'introgression', requires a series of crosses over a number of generations. Without utilizing genetic markers, introgression can be performed only if the genotype of the locus in question can be readily ascertained in the progeny of the different crosses. If the locus being introgressed is linked to a genetic marker, then even alleles without a readily identifiable phenotype can be transferred from one strain to another. Introgression is further facilitated if the QTL is bracketed on either side by segregating genetic markers. If the recombination frequency between the two genetic markers is low, then it can be assumed that double recombinants will be extremely rare, and by selecting on the desired parental genotype, the QTL effect will be transferred nearly intact. Even for genes of major effect, such as the dwarf genes in wheat [17], linkage to segregating genetic markers would allow for earlier identification than possible by genotyping based on quantitative trait expression.

If the cross is made between two lines with a saturated RFLP map, then the process can be accelerated by selecting those progeny with both the desired allele for the locus being transferred, and the greatest number of alleles from the recurrent strain for the other loci [7, 31, 46]. If individuals are selected at random for the background genome, about 99% of the genome $(1 - 0.5^6)$ will be from the recurrent parent after six generations of backcross. By selection on genetic markers, nearly the same level could be reached after only three generations, with 30 individuals genotyped per generation [46].

Conclusions

Statistical methodologies are currently available to detect segregating loci affecting quantitative traits via linkage to RFLPs. The statistical methodology of choice will depend on the experimental design, and the computing resources available. Algorithms based on maximum likelihood are able to make optimum use of all data, but are not available in standard statistical packages, and will generally require more computing resources than those based on least-squares analysis. For QTL linked to a single marker, estimates of recombination frequency can only be derived by ML, and least squares estimated of QTL effects will be biased. Various techniques can be used to increase power of RFLP analysis, either by increasing the number of generations or the number of individuals scored for the quantitative traits. For all experimental designs, RFLP analysis of at least a few hundred individuals will be required to detect segregating QTL responsible for 1–5% of the phenotypic variance.

Acknowledgements

I would like to thank Prof. Morris Soller and Dr. Jaques Beckmann for useful comments. Most of this research was supported by the US-Israel Binational Agricultural Research and Development Fund (BARD).

References

1. Arave CW, Lamb RC, Hines HC (1971) Blood and milk protein polymorphisms in relation to feed efficiency and production traits of dairy cattle. J Dairy Sci 54: 106–112.
2. Asins MJ, Carbonell EA (1988) Detection of linkage between restriction fragment length polymorphisms and quantitative traits. Theor Appl Genet 76: 623–626.
3. Bailey NT (1961) Introduction to the Mathematical Theory of Genetical Linkage. Oxford: Clarendon Press.
4. Beckmann JS, Soller M (1983) Restriction fragment length polymorphisms in genetic improvement: methodologies, mapping and costs. Theor Appl Genet 67: 35–43.
5. Beckmann JS, Soller M (1988) Detection of linkage between marker loci and loci affecting quantitative traits in crosses between segregating populations. Theor Appl Genet 76: 228–236.
6. Beever JS, George PD, Fernando RL, Stormont CJ, Lewin HA (1990) Associations between genetic markers and growth and carcass traits in a paternal half-sib family of Angus cattle. J Anim Sci 68: 337–344.
7. Bernatzky R, Tanksley SD (1986) Toward a saturated linkage map of tomato based on isozymes and random cDNA sequences. Genetics 112: 887–898.
8. Brum EW, Rausch WH, Hines HC, Ludwick TM (1968) Association between milk and blood polymorphism types and lactation traits of Holstein cattle. J Dairy Sci 51: 1031–1038.
9. Cantet RJC, Smith C (1991) Reduced animal model for marker assisted selection using best linear unbiased prediction. Genet Sel Evol (In press).
10. Dahlquist G, Bjorck A (1974) Numerical Methods. Prentice Hall, Englewood Cliffs, NJ.
11. Darvasi A (1990) Analysis of genes affecting a quantitative trait with the aid of bracketing genetic markers, by maximum likelihood methodology. M. Sc. Thesis, Hebrew University, Jerusalem (In Hebrew).
12. Dempster AP, Laird NM, Rubin DB (1977) Maximum likelihood from incomplete data via the EM algorithm. J Roy Stat Soc (Series B) 39: 1–38.
13. Dentine MR, Cowan CM (1990) An analytical model for the estimation of chromosome substitution effects in the offspring of individuals heterozygous at a segregating marker locus. Theor Appl Genet 79: 775–780.
14. Edwards MD, Stuber CW, Wendel JF (1987) Molecular-marker-facilitated investigations of quantitative trait loci in maize. I. Numbers, genomic distribution and types of gene action. Genetics 116: 113–125.
15. Falconer DS (1981) Introduction to Quantitative Genetics, 2nd ed. Longman, New York.
16. Fernando R, Grossman M (1989) Marker assisted selection using best linear unbiased prediction. Genet Sel Evol 21: 467–477.
17. Gale MD, Youssefian S (1985) Dwarfing genes in wheat. In: Russel GE (ed.) Progress in Plant Breeding, pp. 1–35. London: Butterworth.
18. Gelderman H (1975) Investigations on inheritance of quantitative characters in animals by gene markers. I. Methods. Theor Appl Genet 46: 319–330.
19. Gelderman H, Peiper U, Roth B (1985) Effects of marker chromosome sections on milk performance in cattle. Theor Appl Genet 70: 138–146.
20. Goyon DS, Mather RE, Hines HC, Haenlein GFW, Arave CW, Gaunt SN (1987) Associations of bovine blood and milk polymorphisms with lactation traits: Holsteins. J Dairy Sci 70: 2585–2598.
21. Haenlein GFW, Goyon DS, Mather RE, Hines HC (1987) Associations of bovine blood and milk polymorphisms with lactation traits: Guernseys. J Dairy Sci 70: 2599–2609.
22. Hines HC, Kiddy CA, Brum EW, Arave CW (1969) Linkage among cattle blood and milk polymorphisms. Genetics 62: 401–412.
23. Jenson J (1989) Estimation of recombination parameters between a quantitative trait locus (QTL) and two marker gene loci. Theor Appl Genet 78: 613–618.
24. Kahler AL, Wherhahn CF (1986) Associations between quantitative traits and enzyme loci in the F2 population of a maize hybrid. Theor Appl Genet 72: 15–26.

25. Knapp SJ, Bridges WC, Birkes D (1990) Quasi-Mendelian analyses of quantitative trait loci using molecular marker linkage maps. Theor Appl Genet 79: 583–592.
26. Lande R, Thompson R (1990) Efficiency of marker-assisted selection in the improvement of quantitative traits. Genetics 124: 743–756.
27. Lander ES, Botstein D (1989) Mapping Mendelian factors underlying quantitative traits using RFLP linkage maps. Genetics 121: 185–199.
28. Lebowitz RJ, Soller M, Beckmann JS (1987) Trait-based analyses for the detection of linkage between marker loci and quantitative trait loci in crosses between inbred lines. Theor Appl Genet 73: 556–562.
29. Lehmann EL (1983) Theory of Point Estimation. ,New York: John Wiley.
30. Neimann-Sorensen A, Roberson A (1961) The association between blood groups and several production characters in three Danish cattle breeds. Acta Agric Scand 11: 163–196.
31. Paterson AH, Lander ES, Hewitt JD, Peterson S, Lincoln SE, Tanksley SD (1988) Resolution of quantitative traits into Mendelian factors by using a complete linkage map of restriction fragment length polymorphisms. Nature 335: 721–726.
32. Sandler L, Hiraizumi Y, Sandler I (1959) Meiotic drive in natural populations of *Drosophila melanogaster*. I. The cytogenic basis of segregation distortion. Genetics 44: 233–250.
33. SAS Institute, Inc (1985) SAS User's Guide: Statistics, Version 5. SAS Institute, Inc., Cary, NC.
34. Sax K (1923) The association of size differences with seed-coat pattern and pigmentation in *Phaseeolus vulgaris*. Genetics 8: 552–560.
35. Searle SR (1971) Linear Models. New York: John Wiley.
36. Searle SR (1982) Matrix Algebra Useful for Statistics. New York: John Wiley.
37. Simpson SP (1989) Detection of linkage between quantitative trait loci and restriction fragment length polymorphisms using inbred lines. Theor Appl Genet 77: 815–819.
38. Smith C, Simpson SP (1986) The use of genetic polymorphisms in livestock improvement. J Anim Breed Genet 103: 205–217.
39. Smith C, Webb AJ (1981) Effects of major genes on animal breeding strategies. J Anim Breed Genet 98: 161–169.
40. Soller M, Genizi A (1978) The efficiency of experimental designs for the detection of linkage between a marker locus and a locus affecting a quantitative trait in segregating populations. Biometrics 34: 47–55.
41. Soller M, Genizi A, Brody T (1976) On the power of experimental designs for the detection of linkage between marker loci and quantitative loci in crosses between inbred lines. Theor Appl Genet 47: 35–59.
42. Soller M, Beckmann JS (1982) Restriction fragment length polymorphisms and genetic improvement. Proc 2nd World Congr Genet Appl Livest Prod, Madrid 6: 396–404.
43. Soller M, Beckmann JS (1990) Marker-based mapping of quantitative trait loci using replicated progeny. Theor Appl Genet 80: 205–208.
44. Stuber CW, Edwards MD, Wendel JF (1987) Molecular marker-facilitated investigations of quantitative trait loci. II. Factors influencing yield and its component traits. Crop Sci 27: 639–648.
45. Tanksley SD, Medina-Filho H, Rick CM (1982) Use of naturally-occurring enzyme variation to detect and map genes controlling quantitative traits in an interspecific backcross of tomato. Heredity 49: 11–25.
46. Tanksley SD, Yound ND, Paterson AH, Bonierbale MW (1989) RFLP mapping in plant breeding: new tools for an old science. Bio/technology 7: 257–264.
47. Thoday JM (1961) Location of polygenes. Nature 191: 368–370.
48. Thompson JN, Thoday JM (1979) Quantitative Genetic Variation. Academic Press, London.
49. Weller JI (1986) Maximum likelihood techniques for the mapping and analysis of quantitative trait loci with the aid of genetic markers. Biometrics 42: 627–640.
50. Weller JI (1987) Mapping and analysis of quantitative trait loci in *Lycopersicon*. Heredity 59: 413–421.

51. Weller JI, Kashi Y, Soller M (1990) Estimation of sample size necessary for genetic mapping of quantitative traits in dairy cattle using genetic markers. J Dairy Sci 73: 2525–2532.
52. Weller JI, Soller M (1981) Methods for production of multimarker strains. Theor Appl Genet 59: 73–77.
53. Weller JI, Soller M, Brody T (1988) Linkage analysis of quantitative traits in an interspecific cross of tomato (*Lycopersicon esculentum* × *Lycopersicon pimpinellifolium*) by means of genetic markers. Genetics 118: 329–339.
54. Zhuchenko AA, Korol AB, Andryushchenko VK (1979) Linkage between loci of quantitative characters and marker loci. Genetika 14: 771–778.
55. Zhuchenko, AA, Samovol AP, Korol AB, Andryushchenko VK (1979) Linkage between loci of quantitative characters and marker loci. II. Influence of three tomato chromosomes on variability of five quantitative characters in backcross progenies. Genetika 15: 672–683.

10. Mapping quantitative trait loci using nonsimultaneous and simultaneous estimators and hypothesis tests

S.J. KNAPP[1], W.C. BRIDGES[2] & B.-H. LIU[1]

[1]*Department of Crop and Soil Science, OR, USA; State University, Corvallis, OR, USA;* [2]*Department of Experimental Statistics, Clemson University, Clemson, SC, USA*

Introduction

Advances in methods for assaying DNA polymorphisms, e.g., restriction fragment length polymorphisms (RFLPs) and random amplified DNA polymorphisms (RAPDs) [15], have made it feasible to greatly increase the density of linkage maps of many plant species. These maps have revolutionized quantitative genetics by creating the technological base necessary for mapping genes underlying quantitative traits, so-called quantitative trait loci (QTL) [1, 4, 5, 7, 9, 12, 16, see Weller, this book].

Classic quantitative genetic methods are useful for making inferences about population parameters, e.g., genetic variances and heritability. Hypotheses tested by these methods describe the characteristics of populations, but not of genes. Specifically, classical methods do not lead to an understanding of the effects and locations of genes underlying quantitative traits or to the discovery of favorable alleles. This information is essential for practicing marker-assisted selection (MAS) [8, 14, 18, 19].

The abundance of DNA polymorphisms and advances in QTL mapping technology have made the widespread use of MAS feasible [1, 14]. Although the efficiency and economics of MAS are often debated, there are few if any arguments against the value of mapping QTL to gain a deeper understanding of the genetics of a complex trait, and this is perhaps the greatest value of QTL mapping technology to plant breeding. Experiments with maize, tomato, and many other species have begun to unravel the genetics of traits which were formerly poorly understood [2 and many others]. QTL must be mapped before selection can be exerted against marker loci linked to QTL, and how this is done might affect selection efficiency. Towards this end, we review statistical topics which determine the validity of hypothesis tests about QTL and affect the practice of MAS.

Regardless of the kind of progeny used and the complexity of an experimental design, the data for QTL genotypes are unbalanced and, if many loci are simultaneously mapped, then many QTL genotypes may be missing. This has important consequences for mapping QTL and building valid multilocus models of the genetics of QTL. In this paper, we examine the ramifications of unbalanced data and missing cells, especially as they pertain to nonsimulta-

J.S. Beckmann and T.C. Osborn (eds) Plant Genomes: Methods for Genetic and Physical Mapping, 209–237.

neous and simultaneous hypothesis tests. This subject is independent of the method used to map QTL [4, 5, 7, 9, 12, 16, see Weller, this book].

At various points throughout this paper, we make reference to QTL-STAT. QTL-STAT is software for mapping QTL (Liu and Knapp, unpublished). The QTL-STAT prototype was used for the mapping example described in the last section of this paper. The capabilities of QTL-STAT are to be described elsewhere.

Simultaneous and nonsimultaneous mapping

QTL parameters must be simultaneously estimated to make valid inferences about the effects of QTL. This much is obvious. Experimenters do not choose to do otherwise, but population sizes are usually insufficient to simultaneously map QTL without resorting to nonsimultaneous hypothesis tests as a means of selecting QTL for simultaneous hypothesis tests. In principle, this problem does not differ from independent variable selection. The number of QTL which can be simultaneously mapped is a function of the number of individuals or lines in a population (N) and whether or not there are interactions between loci-interactions profoundly affects final model size and the feasibility of simultaneous tests.

Classic problems in unbalanced and missing cell analysis of variance are the crux of the problem of building valid models of the genetics of quantitative trait loci. These problems are not alleviated by maximum likelihood estimation or interval mapping, and are independent of the statistical model used, e.g., linear, nonlinear, or normal distribution mixture models [4, 5, 7, 9, 12, 16, see Weller, this book]. We use linear models throughout this paper, but the principles hold for other statistical models. The problem of handling unbalanced data and missing cells is the major thrust of this paper.

Population size, specifically the number of individuals or lines used, determines the number of parameters which can be simultaneously estimated, and this number is usually far less than the number of observations. If QTL are separately mapped and the outcome is used to propose a genetic model for a trait, then misleading genetic models may emerge, e.g., spurious QTL may be part of the model and important QTL may be excluded from the model. This does not mean that genuine QTL are not part of the model, but that genuine and spurious QTL might make up a particular model, and important QTL may not be part of the model. A reanalysis of many of the experiments reported in the literature thus far should bear this out. It is important to understand this problem because the method used to assemble a model may determine which QTL end up in the final model and, as a consequence, which QTL are used in marker-assisted selection.

How important is it to eliminate spurious QTL from a model? If selection pressure is exerted on spurious QTL, then how does this affect the efficiency and efficacy of marker-assisted selection? These are important questions. The answers are determined by mapping objectives and the cost of false positives and negatives. Putting selection pressure on a neutral locus – spurious QTL – basically means selection pressure is not being exerted against the trait, because it makes no difference which marker allele is selected if the marker locus is not linked to an important QTL. But it may dramatically affect selection intensity and effective population size. These factors decrease the rate of fixation of favorable alleles and increase the probability of fixation and loss of alleles as a consequence of random drift. Adding loci to a selection index decreases the pool of favorable allele homozygotes, making it more difficult to find individuals homozygous for favorable alleles – it decreases the overall frequency of favorable alleles. So the cost for putting selection pressure on spurious QTL is decreased selection efficiency. The exact cost is variable and debatable, but it is always important to try to build valid models of the genetics of QTL, even if the only mapping objective is selecting QTL for MAS. The cost of false positives may be far greater if there are other mapping objectives.

Unbalanced data and missing cells

To review the theory behind building valid multilocus QTL models, we use two-locus genetic models and proceed as if the QTL genotypes are known. In practice, of course, they are not known and, if linear or mixture models are used, must be estimated [4, 5, 7, 9, 12, 16, see Weller this book]. More specifically, the posterior probabilities of individuals belonging to QTL genotypic classes must be estimated. This increases estimation complexity, but the consequences of unbalanced data and missing cells are the same whether or not the genotypes are known with certainty. The problem in estimating QTL genotypic values does not arise when using nonlinear models to map QTL because marker genotypes *per se* are used as independent variables [5, 7]; however, the problem of unbalanced data and missing cells is still pertinent.

Suppose a model is fit for two QTL, Q_1 and Q_2, which are segregating in an F_2 population. Further suppose observations are made on unreplicated F_2 individuals, and QTL genotypes are laid out in a completely randomized experimental design. A suitable linear model for the two-locus model with interaction is

$$y_{ijk} = \mu + \lambda_i + \gamma_j + \lambda\gamma_{ij} + e_{ijk} \tag{1}$$

where $i = 1, 2, \ldots, s$, $j = 1, 2, \ldots, t$, $k = 1, 2, \ldots, n_{ij}$, i indexes genotypes at quantitative trait locus Q_1, j indexes genotypes at quantitative trait locus Q_2, k indexes the number of observations of the ijth QTL genotype, n_{ij} is the number of observations of the ijth QTL genotype, s is the number of genotypes at the Q_1 locus, t is the number of genotypes at the Q_2 locus, y_{ijk} is the ijkth observation of the quantitative trait, μ is the population mean, λ_i is the effect of the ith genotype at Q_1, γ_j is the effect of the jth genotype at Q_2, $\lambda\gamma_{ij}$ is the effect of the interaction between the *i*th and *j*th genotypes at Q_1 and Q_2, respectively.

Another model we need to examine is the two-locus model without interaction between Q_1 and Q_2. A suitable linear model is

$$y_{ijk} = \mu + \lambda_i + \gamma_j + e_{ijk}. \tag{2}$$

This is just model (1) with the interaction between Q_1 and Q_2 dropped.

The 'factors' in models (1) and (2) are quantitative trait loci Q_1 and Q_2. Model (1) is a two-factor model with interaction [11]. Model (2) is a two-factor model without interaction. The 'factor levels' are the QTL genotypes at Q_1 and Q_2 [11]. If the number of observations or replications of QTL genotypes are unequal, then the n_{ij} within each cell (the number of progeny of a particular QTL genotype) are not equal and the data are 'unbalanced'.

The data for every progeny type are unbalanced to some extent, but the extent of unbalance is different for different progeny types and increases as the number of loci increases and the number of progeny in the population decreases. Doubled haploid and recombinant inbred progeny data, for example, are obviously less unbalanced than F_2 or F_3 progeny data. The expected number of observations within 'cells' is determined by the expected segregation ratio and is equal to the number of individuals or lines of a specific QTL genotype. Cells for models (1) and (2) are QTL genotypes.

The number of observations for each QTL genotype in an F_2 population segregating for many QTL is going to be very different, and many QTL genotypes will be missing. When this happens, the data have 'missing cells'.

A model is 'balanced' when each cell has an equal number of observations. Balanced models have the nice property of orthogonality among independent variables or factors, e.g., dummy variables indexing QTL genotypes. Unbalanced models have the property of nonorthogonality among QTL genotypes. When the data are unbalanced and two or more QTL are in the model, there is more than one way to estimate the sums of squares and to test hypotheses about model effects [11]. Under these circumstances, caution is needed to avoid biased estimates of QTL genotype means and biased tests which may not address stated hypotheses. Whether or not an estimator or test statistic is biased is determined by the sums of squares used and the presence of missing cells and interactions between QTL (Fig. 1). Some of the effects of the interaction model (1), for example, are not estimable when cells are missing.

Biased and unbiased estimates of QTL genotype means

Unbiased estimates of the means of QTL genotypes are needed to make valid inferences about QTL. Getting unbiased estimates from unbalanced data and missing cells requires more care than is necessary with balanced data. Our review of this subject is geared towards ensuring that statistical bias does not

	No Missing Cells	Missing Cells
Multilocus Model Without Interaction	Estimable and Unbiased	Estimable and Unbiased
Multilocus Model With Interaction	Estimable and Unbiased	Unestimable and Biased

Fig. 1. Estimability and bias of least-square means for different statistical models and data structures.

interfere with making accurate inferences about QTL. We take a conservative stance towards this problem. If unbiased estimators are used and statistical bias is not serious for a particular trait or population, then no harm is done; however, if they are not used and bias is serious, then erroneous inferences about the effects of QTL might be made. Statistical bias might be negligible for a particular trait or population, but this is not universally true. Furthermore, there is no foolproof way of knowing if bias is significant for a given mapping problem.

The 'marginal mean' of a QTL genotype is the sum of the observations of that QTL genotype divided by the number of observations, e.g., the marginal mean of genotype 1 at Q_1 is

$$\bar{y}_{1..} = \frac{\sum_{j=1}^{t} \sum_{k=1}^{n_{ij}} y_{1jk}}{\sum_{j} n_{1j}}.$$

This estimator is biased when the data are unbalanced.

Likewise, differences between marginal means are biased estimators of the effects of QTL genotypes when the data are unbalanced. These estimators are biased because they depend on the cell frequencies (number of observations per genotype). Means of Q_1 locus genotypes are affected by the sample of genotypes observed at the Q_2 locus and vice versa. The magnitude of the bias is determined by the size of the effects of the QTL and the extent of the unbalance. The latter is determined by sample size, the number of quantitative trait loci, and the observed QTL genotype frequencies.

The 'least square mean' of a QTL genotype is the sum of the cell means for that genotype divided by the number of cells where the cell means are $\bar{\mu}_{ij} = \bar{y}_{ij.}$, e.g., the least square mean of genotype 1 at Q_1 is

$$\bar{\mu}_{1..} = \frac{\sum_{j=1}^{t} \bar{\mu}_{1j.}}{t}.$$

Least-square means and differences between least-squares means are unbiased because they do not depend on cell frequencies [3, 10]. Least-square means and marginal means are only equal when the data are balanced or proportionally unbalanced [3].

It is important to understand how bias arises, and how this affects the validity of inferences about QTL. If a QTL is mapped and other important QTL are not simultaneously mapped, then hypothesis tests and estimates of QTL means of the mapped QTL are biased. The direction and magnitude of the bias is determined by the effects of QTL excluded from the model and the sample of QTL genotypes in the population. If every important QTL is simultaneously mapped, then unbiased hypothesis tests and mean estimators exist for certain models (Fig. 1). Contradictory inferences about the genetics of QTL often emerge from nonsimultaneous and simultaneous hypothesis tests because nonsimultaneous tests lead to rejecting or failing to reject null hypotheses as a consequence of 'sampling bias'. Sampling bias is caused by unbalanced samples of genotypes in a population.

To show this, we use a two-locus F_2 example without interaction. We used observations for 19 F_2 progeny segregating for two independent QTL, Q_1 and Q_2 (Table 1). Model 2 is appropriate for these data. The expected values of these observations are listed for model 2 (Table 2). We used a small 'example population' to illustrate salient points about bias and to let us show the expected values of the observations. Later we examine how increasing sample size affects bias. The observed frequency of genotypes at Q_1 and Q_2 is close to the expected frequency for two independent loci, so these data are close to being proportionally unbalanced (Table 2).

A typical objective of an F_2 mapping experiment is to estimate the additive and dominance effects of QTL for every marked segment of the genome, and to test whether or not these effects are significantly different from zero. The additive and dominance effects of QTL are customarily defined as differences

Table 1. An unbalanced data example for an F_2 population segregating for two quantitative trait loci Q_1 and Q_2. The units of the quantitative trait are arbitrary. Genotypes 1 and 3 are the homozygotes for each locus. Genotype 2 is the heterozygote for each locus

Q_1 genotype	Q_2 genotype		
	1	2	3
1	10	15, 16, 16	20, 21
2	11, 12	16, 17, 17, 18	21, 22
3	12, 13	18, 19	23

Table 2. Effects for model (2), the two-factor model without interaction, for the example data of Table 1. The effects of model (2) are defined in the text. The factors of the model are quantitative trait loci Q_1 and Q_2 segregating in an F_2 population. Genotypes 1 and 3 are the homozygotes for each locus. Genotype 2 is the heterozygote for each locus

Q_1 genotype	Q_2 genotype		
	1	2	3
1	$\mu + \lambda_1 + \gamma_1 + e_{111}$	$\mu + \lambda_1 + \gamma_2 + e_{121}$ $\mu + \lambda_1 + \gamma_2 + e_{122}$ $\mu + \lambda_1 + \gamma_2 + e_{123}$	$\mu + \lambda_1 + \gamma_3 + e_{131}$ $\mu + \lambda_1 + \gamma_3 + e_{132}$
2	$\mu + \lambda_2 + \gamma_1 + e_{211}$ $\mu + \lambda_2 + \gamma_1 + e_{212}$	$\mu + \lambda_2 + \gamma_2 + e_{221}$ $\mu + \lambda_2 + \gamma_2 + e_{222}$ $\mu + \lambda_2 + \gamma_2 + e_{223}$ $\mu + \lambda_2 + \gamma_2 + e_{224}$	$\mu + \lambda_2 + \gamma_3 + e_{231}$ $\mu + \lambda_2 + \gamma_3 + e_{232}$
3	$\mu + \lambda_3 + \gamma_1 + e_{311}$ $\mu + \lambda_3 + \gamma_1 + e_{312}$	$\mu + \lambda_3 + \gamma_2 + e_{321}$ $\mu + \lambda_3 + \gamma_2 + e_{322}$	$\mu + \lambda_3 + \gamma_3 + e_{331}$

between means of QTL genotypes. Let μ_1 and μ_3 be means of homozygous genotypes at Q_1 and μ_2 be the mean of the heterozygous genotype at Q_1, then the additive and dominance effects of Q_1 are $\mu_1 - \mu_3$ and $\mu_1 + \mu_3 - 2\mu_2$, respectively. The additive and dominance effects of QTL are frequently estimated by taking differences between marginal means. The difference between marginal means of Q_1 locus homozygotes, expressed as estimates of model (2) effects (Table 2), is

$$\bar{y}_{1..} - \bar{y}_{3..} = \frac{1}{6} [(\hat{\lambda}_1 + \hat{\gamma}_1) + (\hat{\lambda}_1 + \hat{\gamma}_2) + (\hat{\lambda}_1 + \hat{\gamma}_2) + (\hat{\lambda}_1 + \hat{\gamma}_2) + (\hat{\lambda}_1 + \hat{\gamma}_3)$$

$$+ (\hat{\lambda}_1 + \hat{\gamma}_3)] - \frac{1}{5} [(\hat{\lambda}_3 + \hat{\gamma}_1) + (\hat{\lambda}_3 + \hat{\gamma}_1) + (\hat{\lambda}_3 + \hat{\gamma}_2) + (\hat{\lambda}_3 + \hat{\gamma}_2) + (\hat{\lambda}_3 + \hat{\gamma}_3)]$$

$$= (\hat{\lambda}_1 - \hat{\lambda}_3) + \left[\left(\frac{1}{6} - \frac{2}{5} \right) \hat{\gamma}_1 + \left(\frac{1}{2} - \frac{2}{5} \right) \hat{\gamma}_2 + \left(\frac{1}{3} - \frac{1}{5} \right) \hat{\gamma}_3 \right]$$

$$= (\hat{\lambda}_1 - \hat{\lambda}_3) + [-0.23 \hat{\gamma}_1 + 0.10 \hat{\gamma}_2 + 0.13 \hat{\gamma}_3]; \tag{3}$$

hence, (3) is a biased estimator of the additive effect of Q_1. The bias is $-0.23\gamma_1 + 0.10\gamma_2 + 0.13\gamma_3$. Likewise, the marginal mean estimator of the dominance effect of Q_1 is

$$\bar{y}_{1..} + \bar{y}_{3..} - 2\bar{y}_{2..} = \frac{1}{6} [(\hat{\lambda}_1 + \hat{\gamma}_1) + 3(\hat{\lambda}_1 + \hat{\gamma}_2) + 2(\hat{\lambda}_1 + \hat{\gamma}_3)]$$

$$+ \frac{1}{5} \left[2(\hat{\lambda}_3 + \hat{\gamma}_1) + 2(\hat{\lambda}_3 + \hat{\gamma}_2) + (\hat{\lambda}_3 + \hat{\gamma}_3) \right]$$

$$- 2\left\{ \frac{1}{8} \left[2(\hat{\lambda}_2 + \hat{\gamma}_1) + 4(\hat{\lambda}_2 + \hat{\gamma}_2) + 2(\hat{\lambda}_2 + \hat{\gamma}_3) \right] \right\}$$

$$= (\hat{\lambda}_1 + \hat{\lambda}_3 - 2\hat{\lambda}_2) + [0.067\,\hat{\gamma}_1 - 0.1\,\hat{\gamma}_2 + 0.033\,\hat{\gamma}_3]; \tag{4}$$

hence, (4) is a biased estimator of the dominance effect of Q_1. The bias is $0.067\,\hat{\gamma}_1 - 0.1\,\hat{\gamma}_2 + 0.033\,\hat{\gamma}_3$.

Estimators (3) and (4) are biased because the sample of genotypes at Q_1 and Q_2 is not balanced and, as a consequence, the effects of the Q_2 locus do not 'cancel' as they do when the data are balanced. The bias of estimators (3) and (4) for the two-locus model without interaction or model (2) is $k_1\,\hat{\gamma}_1 + k_2\,\hat{\gamma}_2 + k_3\,\hat{\gamma}_3$. Values of the coefficients k_1, k_2, and k_3 are determined by the array of QTL genotypes or the cell frequencies. These coefficients are equal to zero when the data are balanced or proportionally unbalanced, otherwise they are not equal to zero [3].

It is not necessary to use marginal means as estimators of QTL genotype means because, as we show below, least-square means, which are unbiased estimators, can be used; however, we defined the bias of marginal mean estimators (3) and (4) to make a crucial point. These estimators are biased when Q_1 and Q_2 are nonsimultaneously or simultaneously mapped. If they are nonsimultaneously mapped, then there is no mechanism for getting unbiased estimates and hypothesis tests. The only estimators which exist for non-simultaneous hypothesis tests are differences between marginal means, and these are biased. If they are simultaneously mapped, then there is a mechanism for getting unbiased estimates and hypothesis tests. This is a more serious problem for QTL which have minor effects segregating with QTL which have major effects because differences between means of QTL genotypes at minor QTL may be overwhelmed by differences between means of QTL genotypes at major QTL.

We reviewed the bias problem because it shows why many of the QTL effects found by nonsimultaneous hypothesis tests might be erroneous, and why many important QTL effects might not be found. Inferences made from biased hypothesis tests are often erroneous, and many QTL found by nonsimul-taneous tests are false positives caused by sampling bias. Many of these can be eliminated by subsequent simultaneous hypothesis tests. But a more nettle-some problem are false negatives caused by sampling bias. False negatives are QTL which underlie a trait, but which are not found because of sampling bias. The two-locus F_2 example (Tables 1 and 2) shows this and demonstrates why it can be misleading to make interences from nonsimultaneous hypothesis tests and how important it is to eliminate sampling bias by using simultaneous hypothesis tests.

The bias problem is alleviated by using least-square means [3, 10]; however, the only way to get estimates of least-square means is by simultaneously estimating the parameters of important QTL. We stress important QTL because the size of the effect of a QTL determines its significance as a source of bias. Why this is so should become clear as we work through the example. For the original F_2 example data (Tables 1 and 2) and the two-locus model without interaction, the difference between least-square means for Q_1 locus homozygotes is

$$\bar{\mu}_{1.} - \bar{\mu}_{3.} = \frac{1}{3} \{(\hat{\lambda}_1 + \hat{\gamma}_1) + [(\hat{\lambda}_1 + \hat{\gamma}_2) + (\hat{\lambda}_1 + \hat{\gamma}_2) + (\hat{\lambda}_1 + \hat{\gamma}_2)]/3$$

$$+ [(\hat{\lambda}_1 + \hat{\gamma}_3) + (\hat{\lambda}_1 + \hat{\gamma}_3)]/2\} - \frac{1}{3} \{[(\hat{\lambda}_3 + \hat{\gamma}_1) + (\hat{\lambda}_3 + \hat{\gamma}_1)]/2$$

$$+ [(\hat{\lambda}_3 + \hat{\gamma}_2) + (\hat{\lambda}_3 + \hat{\gamma}_2)]/2 + (\hat{\lambda}_3 + \hat{\gamma}_3)\}$$

$$= \hat{\lambda}_1 + \frac{1}{3}(\hat{\gamma}_1 + \hat{\gamma}_2 + \hat{\gamma}_3) - \hat{\lambda}_3 - \frac{1}{3}(\hat{\gamma}_1 + \hat{\gamma}_2 + \hat{\gamma}_3) = \hat{\lambda}_1 - \hat{\lambda}_3. \tag{5}$$

This is an unbiased estimator of the additive effect of the Q_1 locus. Estimator (5) shows the least-square means are free of the effects of Q_2 and the cell frequencies [3]. The least-square mean estimator of the dominance effect Q_1 is

$$\bar{\mu}_{1.} + \bar{\mu}_{3.} - 2\bar{\mu}_{2.} = \frac{1}{3} \{(\hat{\lambda}_1 + \hat{\gamma}_1) + 3[(\hat{\lambda}_1 + \hat{\gamma}_2)/3] + 2[(\hat{\lambda}_1 + \hat{\gamma}_3)/2]\}$$

$$+ \frac{1}{3} \{2[(\hat{\lambda}_3 + \hat{\gamma}_1)/2] + 2[(\hat{\lambda}_3 + \hat{\gamma}_2)/2] + (\hat{\lambda}_3 + \hat{\gamma}_3)\}$$

$$- 2\frac{1}{3} \{2[(\hat{\lambda}_3 + \gamma_1)/2] + 4[(\hat{\lambda}_3 + \hat{\gamma}_2)/4] + 2[(\hat{\lambda}_3 + \hat{\gamma}_3)/2]\}$$

$$= \hat{\lambda}_1 + \frac{1}{3}(\hat{\gamma}_1 + \hat{\gamma}_2 + \hat{\gamma}_3) + \hat{\lambda}_3 + \frac{1}{3}(\hat{\gamma}_1 + \hat{\gamma}_2 + \hat{\gamma}_3)$$

$$- 2\hat{\lambda}_2 - 2[\frac{1}{3}(\hat{\gamma}_1 + \hat{\gamma}_2 + \hat{\gamma}_3)]$$

$$= \hat{\lambda}_1 + \hat{\lambda}_3 - 2\hat{\lambda}_2; \tag{6}$$

thus, (6) is unbiased as well.

The least-square means themselves are unbiased only when the effects of every important QTL are simultaneously estimated. Suppose, for example, that another independent QTL, say Q_3, is segregating and we used the two-locus model to map Q_1 and Q_2, then the effect of Q_3 might seriously bias the marginal means and he two-locus model may not equal the least-square means for the three-locus model. And so on and so forth.

It might be more instructive for us to show a numerical example using the two-locus F_2 data we fabricated (Table 1). Estimates of the marginal means for genotypes 1, 2, and 3 at Q_1 are

$$\bar{y}_{1..} = \frac{10 + 15 + 16 + 20 + 21}{6} = 16.33,$$

$$\bar{y}_{2..} = \frac{11 + 12 + 16 + 17 + 17 + 18 + 21 + 22}{8} = 16.75,$$

and

$$\bar{y}_{3..} = \frac{12 + 13 + 18 + 19 + 23}{5} = 17.00,$$

respectively. Estimates of the least-square means for genotypes 1, 2, and 3 at QS_1 are

$$\bar{\mu}_{1.} = \frac{10 + (15 + 16 + 16)/3 + (20 + 21)/2}{3} = 15.39,$$

$$\bar{\mu}_{2.} = \frac{(11 + 12)/2 + (16 + 17 + 17 + 18)/4 + (21 + 22)/2}{3} = 16.67,$$

and

$$\bar{\mu}_{3.} = \frac{(12 + 13)/2 + (18 + 19)/2 + 23}{3} = 18.00,$$

respectively. Differences between the estimates of marginal and least-square means are explained by sampling bias.

The additive effect of Q_1 estimated using differences between marginal means and differences between least-square means is $\bar{y}_{1..} - \bar{y}_{3..} = 16.33 - 17.00 = -0.67$ and $\bar{\mu}_{1.} - \bar{\mu}_{3.} = 15.39 - 18.00 = -2.61$, respectively. The difference between marginal means is biased by 1.94 units. This shows how QTL which have major effects seriously bias the effects of QTL which have minor effects. If Q_2 was not in the model, then the difference between least-square means could not be used to eliminate sampling bias and we would erroneously infer that Q_1 has no additive effect – a false negative or Type II error would have been committed.

The dominance effect of Q_1 estimated using marginal and least-square means is $16.33 + 17.00 - 2(16.75) = -0.17$ and $15.39 + 18.00 - 2(16.67) = 0.05$, respectively. The dominance effect of this locus is not great, but the difference between marginal means is nevertheless biased by the sample of genotypes at the Q_2 locus. This example shows sampling bias is bidirectional. Sampling bias leads to Type I and Type II errors. A 'Type II error' is made when a null hypothesis is false but is not rejected. For the F_2 example, a Type I error was made when the hypothesis of no additive effect was tested, whereas a Type II error was made when the hypothesis of no dominance effect was tested; thus, for the Q_1 locus, the additive and dominance effects are biased in different directions.

Earlier we mentioned proportionally unbalanced n_{ij}. This condition arises in F_2 progeny when the observed genotype or cell frequencies are equal to the expected genotype frequencies for independent QTL. Suppose observations are taken from 16 F_2 progeny segregating for two QTL, Q_1 and Q_2, and the observed number of observations for each QTL genotype equals the expected number (Table 3). When the n_{ij} are proportionately unbalanced, differences between marginal means are unbiased estimators of additive and dominance effects of QTL; thus,

$$\bar{y}_{1..} - \bar{y}_{3..} = \frac{1}{4} [(\lambda_1 + \gamma_1) + 2(\lambda_1 + \gamma_2) + (\lambda_1 + \gamma_3)]$$

$$- \frac{1}{4} [(\lambda_3 + \gamma_1) + 2(\lambda_3 + \gamma_2) + (\lambda_3 + \gamma_3)]$$

$$= (\lambda_1 - \lambda_3)$$

Table 3. Effects for model (2), the two-factor model without interaction for proportionally unbalanced n_{ij}. The effects of model (2) are defined in the text. The factors of the model are quantitative trait loci Q_1 and Q_2 segregating in an F_2 population. There are 16 observations. The number of observations within each cell is equal to the expected segregation ratio. Genotypes 1 and 3 are the homozygotes for each locus. Genotype 2 is the heterozygote for each locus

Q_1 genotype	Q_2 genotype		
	1	2	3
1	$\mu + \lambda_1 + \gamma_1 + e_{111}$	$\mu + \lambda_1 + \gamma_2 + e_{121}$ $\mu + \lambda_1 + \gamma_2 + e_{122}$	$\mu + \lambda_1 + \gamma_3 + e_{131}$
2	$\mu + \lambda_2 + \gamma_1 + e_{211}$ $\mu + \lambda_2 + \gamma_1 + e_{212}$	$\mu + \lambda_2 + \gamma_2 + e_{221}$ $\mu + \lambda_2 + \gamma_2 + e_{222}$ $\mu + \lambda_2 + \gamma_2 + e_{223}$ $\mu + \lambda_2 + \gamma_2 + e_{224}$	$\mu + \lambda_2 + \gamma_3 + e_{231}$ $\mu + \lambda_2 + \gamma_3 + e_{232}$
3	$\mu + \lambda_3 + \gamma_1 + e_{311}$	$\mu + \lambda_3 + \gamma_2 + e_{321}$ $\mu + \lambda_3 + \gamma_2 + e_{322}$	$\mu + \lambda_3 + \gamma_3 + e_{331}$

and

$$\bar{y}_{1..} + \bar{y}_{3..} - 2\bar{y}_{2..} = \frac{1}{4} \left[(\lambda_1 + \gamma_1) + 2(\lambda_1 + \gamma_2) + (\lambda_1 + \gamma_3) \right]$$

$$+ \frac{1}{4} \left[(\lambda_3 + \gamma_1) + 2(\lambda_3 + \gamma_2) + (\lambda_3 + \gamma_3) \right] - 2\frac{1}{8} \left[2(\lambda_2 + \gamma_1) \right.$$

$$+ 4(\lambda_2 + \gamma_2) + 2(\lambda_2 + \gamma_3)]$$

$$= (\lambda_1 + \lambda_3 - 2\lambda_2),$$

which are unbiased estimators of the additive and dominance effects of Q_1, respectively. Although proportionally unbalanced data rarely arise in practice, the data for many progeny types become more proportionately unbalanced as sample size increases.

Biased and unbiased hypothesis tests about differences between QTL genotype means

Just as there are biased and unbiased estimators of means for unbalanced models, there are biased and unbiased statistics for testing hypotheses about differences between means of QTL genotypes. To be specific, there are four methods for estimating sums of squares (SS) [3]. Nomenclature for these methods is Type I, II, III, and IV SS [3]. Hypothesis tests based on these SS are defined as Type I, II, III, or IV hypothesis tests. The type of sums of squares used determines the nature of the hypothesis being tested. Some are biased and some are not. Type I SS and hypothesis tests for unbalanced data and missing cells are biased. This raises another crucial point about simultaneous hypothesis tests – inferences made from separate nonsimultaneous hypothesis tests use Type I SS and are biased when there are two or more QTL underlying a trait. If a hypothesis is tested about a QTL and other important QTL are not in the model, then there is no mechanism for eliminating the sampling bias arising from QTL excluded from the model. As we stated earlier, this leads to Type I and Type II errors. This problem is equivalent to the problem of getting unbiased estimates of QTL genotype means.

Type I, II, III, and IV sums of squares are equal when the data are balanced, but they are not necessarily equal for unbalanced data and missing cells [3]. These equalities do not hold for every factor in a model either. Although each model must be examined to define its characteristics, models for QTL mapping problems have consistent features. We examine these features using the two-locus example and explain how to get unbiased test statistics and hypothesis tests.

For the two-locus F_2 example, there are at two equally logical ways of getting

Table 4. Types I and III sums of squares (SS) for model (1). The effects of model (1) are defined in the text. Model factors are quantitative trait loci Q_1 and Q_2. Type I SS are shown for fitting Q_1 before Q_2 ($Q_1|Q_2$) or Q_2 before Q_1 ($Q_2|Q_1$). The order is inconsequential for Type III SS

Source	Type I ($Q_1|Q_2$)	Type I ($Q_2|Q_1$)	Type III
Q_1	$R[\lambda|\mu]$	$R[\lambda|\mu, \gamma]$	$R[\lambda|\mu, \gamma, \lambda\gamma]$
Q_2	$R[\gamma|\mu, \gamma]$	$R[\gamma|\mu]$	$R[\gamma|\mu, \lambda, \lambda\gamma]$
$Q_1 \times Q_2$	$R[\lambda\gamma|\mu, \lambda, \gamma]$	$R[\lambda\gamma|\mu, \lambda, \gamma]$	$R[\lambda\gamma|\mu, \lambda, \gamma]$

Type I SS: fitting Q_1 before Q_2 or fitting Q_2 before Q_1. The interaction is normally fit after fitting Q_1 and Q_2. If Q_1 is fit before Q_2 and $Q_1 \times Q_2$, then the reduction SS is $R[\lambda|\mu]$ (Table 4). $R[\lambda|\mu]$ is the reduction SS as a consequence of fitting the effect of $Q_1(\lambda)$ after the population mean (μ). This notation is widely used to differentiate between different SS types [3, 11]. The reduction SS for fitting Q_2 after Q_1 is $R[\gamma|\mu, \lambda]$. Likewise, the reduction SS for fitting the interaction last is $R[\lambda\gamma|\mu, \lambda, \gamma]$ (Table 4). If Q_2 is fit before Q_1 and $Q_1 \times Q_2$, then the reduction SS is $R[\gamma|\mu]$. The reduction SS for fitting Q_1 after Q_2 is $R[\lambda|\mu, \gamma]$. Likewise, the reduction SS for fitting the interaction last is $R[\lambda\gamma|\mu, \lambda, \gamma]$. The different Type I SS for Q_1 and Q_2 are not equal, nor are the hypotheses tested by every SS unbiased (Table 4). This illustrates the problem of having more equations than unknowns. The meaning of a Type I hypothesis test is order-dependent. As a rule, Type I SS and hypothesis tests are biased because they are functions of the cell frequencies [3, 10].

Another way of estimating SS for unbalanced data is to fit each effect after fitting the other effects. This leads to the Type III SS for the two-locus model with interaction (Table 4). Type I and III SS for certain effects are equal, whereas others are not. For the two-locus model with interaction and no missing cells, for example, the Type I and III SS for Q_1 and Q_2 are not equal – $E[\lambda|\mu] \neq R[\lambda|\mu, \gamma] \neq R[\lambda|\mu, \gamma, \lambda\gamma]$ and $R[\gamma|\mu] \neq R[\gamma|\mu, \lambda]$ $\neq R[\gamma|\mu, \lambda, \lambda\gamma]$ – whereas the Type I and III SS for the interaction between Q_1 and Q_2 are equal (Table 4). The Type III SS are not biased by cell frequencies and lead to unbiased tests of hypotheses about differences between means of QTL genotypes [3, 10, 11]. Estimating Type III SS is straightforward, but the effects of every important QTL must be simultaneously estimated to eliminate sampling bias. This is so because the effects of QTL excluded from a model are unestimable, but their effects do not vanish! At the very least, the effect of major QTL should be simultaneously estimated since these are the greatest source of bias. QTL with minor effects are minor sources of bias.

There are only 19 progeny in the F_2 example data (Table 1), which is clearly an unsatisfactory sample size; however, increasing sample size does not decrease the bias problem unless it decreases the extent of unbalance. This is easily seen by multiplying the number of observations within each cell by a factor of 10. Although this increases sample size from 19 to 190, it does not

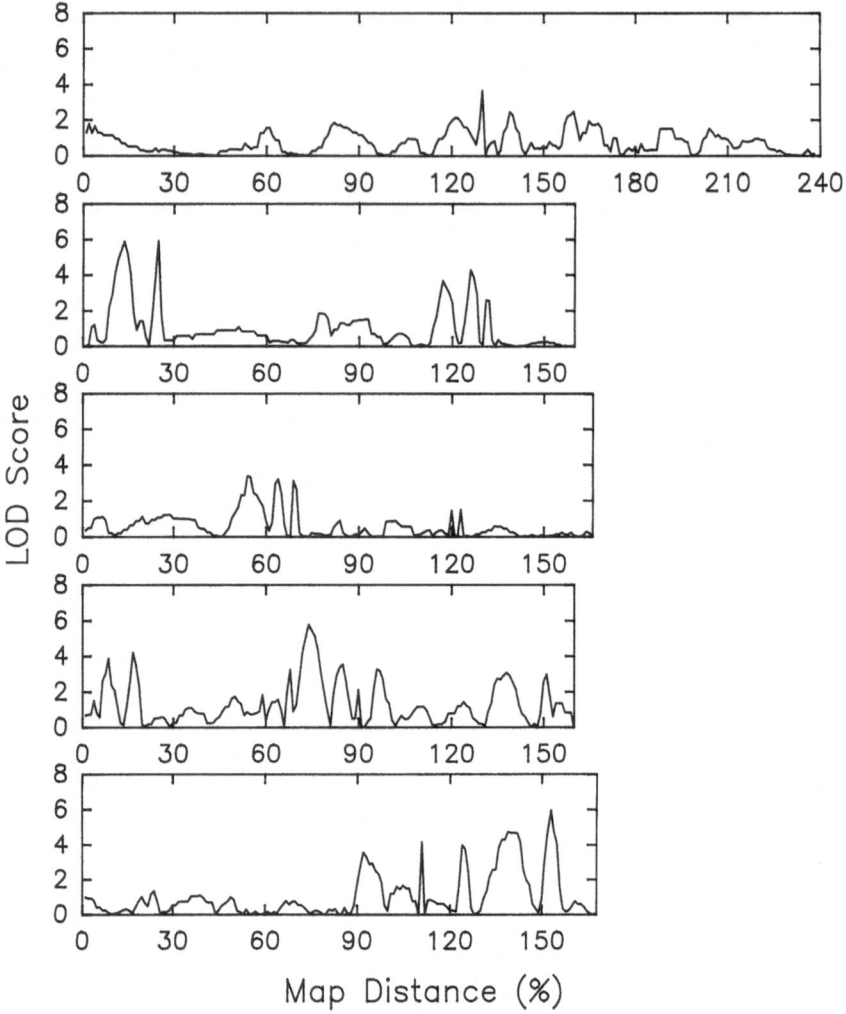

Fig. 2. LOD statistics generated by QTL-STAT by interval mapping of the genome of maize for QTL affecting telomere length (kb). Two replications of 41 recombinant inbred lines were used. Plots are shown for chromosomes 1 through 5 in order with chromosome 1 at the top.

affect the bias coefficients of (3) and (4) whatsoever; however, increasing sample size decreases the extent of unbalance. But the extent of unbalance increases as the number of loci increases. Imagine, for example, the extent of unbalance when three or more QTL are segregating in an F_2 population of 100 to 200 individuals. The sample of genotypes for a given locus might be nearly proportionally unbalanced or balanced, but the sample for several loci is likely to be much more unbalanced.

Multilocus models with missing cells

Up to now, we have confined this review to models where there are no missing QTL genotypes. Missing genotypes obviously arise in practice when the number of loci is great or the number of progeny is not very great. Although we have argued strongly for avoiding the pitfalls of nonsimultaneous hypothesis tests, simultaneous tests have serious pitfalls if there are missing QTL genotypes and interactions between QTL. This is one of the four classes of problems we review in this paper (Fig. 1). The other three are for models with no missing cells or, if there are missing cells, no interactions between QTL. For these three problems, the least-square means are estimable and unbiased (Fig. 3).

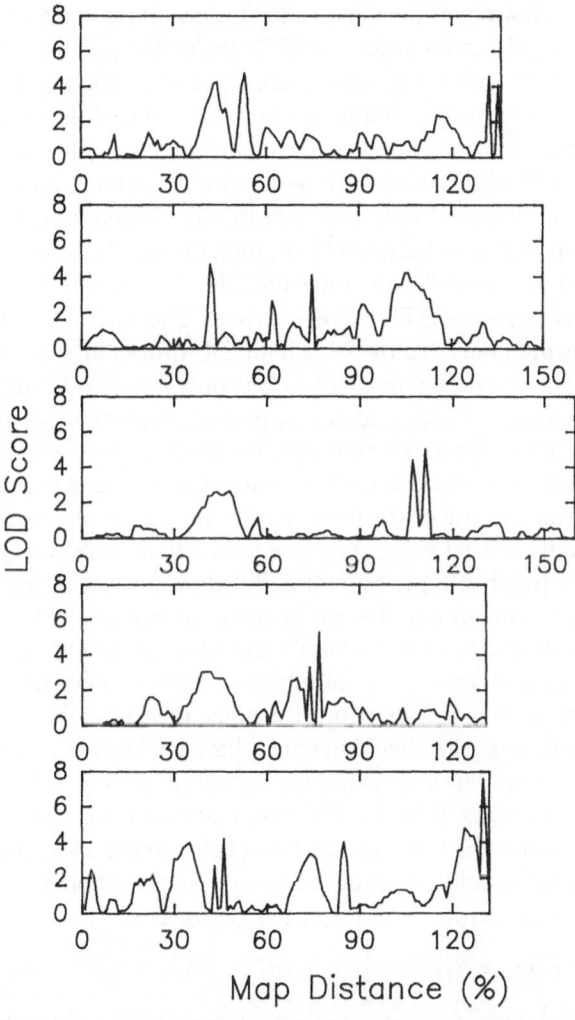

Fig. 3. LOD statistics generated by QTL-STAT by interval mapping of the genome of maize for QTL affecting telomere length (kb). Two replications of 41 recombinant inbred lines were used. Plots are shown for chromosomes 6 through 10 in order with chromosome 6 at the top.

Table 5. A missing cell data example for an F_2 population segregating for two quantitative trait loci Q_1 and Q_2. The units of the quantitative trait are arbitrary. Genotypes 1 and 3 are the homozygotes for each locus. Genotype 2 is the heterozygote for each locus

Q_1 genotype	Q_2 genotype		
	1	2	3
1		15, 16, 16	20, 21
2	11, 12	16, 17, 17, 18	21, 22
3	12, 13	18, 19	

As long as there are no interactions among the QTL, then least-square means are unbiased estimators of the means of QTL genotypes, and Type III tests of hypotheses about the main effects of QTL are unbiased, even when there are many missing QTL genotypes among the 2^k or 3^k genotypes in a population where k is the number of quantitative trait loci. To prove this, we use the original two-locus F_2 example data, but drop the double homozygotes. This reduces the data from 19 to 17 observations. To avoid confusing these data with the data of the original unbalanced data, we rewrote the observations (Table 5).

If there are interactions between QTL, then the least-square means for QTL genotypes with missing cells are 'not estimable'. This is why the assumption of no interactions between QTL is very strong. The interaction problem with missing cells is not alleviated by using a model without interaction or by using nonsimultaneous tests of main effects – it is only ignored. If interactions exist and a two-locus model with interaction is used, then many of the means are not estimable. This repeats an argument we made earlier. A QTL effect, be it a main or interaction effect, does not vanish because it is excluded from the model. These effects still perturb the effects of loci in the model. If the no-interaction model is used when there are interactions between QTL, then it is mathematically feasible to get estimates, but they are biased and meaningless. Missing cells are a major pitfall when interactions between QTL are important because many means are not estimable and meaningful hypothesis tests may not be found. A fractional factorial might be used to solve this problem, but a suitable fraction may not exist or, if it does, then it might be hard to find.

If there are missing cells, then differences between least-square means for the two-locus model without interaction are unbiased as long as there is no interaction between the QTL (Fig. 1). The marginal means for genotypes 1, 2, and 3 at the Q_1 locus for the missing cell data (Table 5) are 17.6, 16.75, and 15.5, respectively, and the additive effect estimated by the difference between marginal means for genotypes 1 and 3 at the Q_1 locus is

$$\bar{y}_{1..} - \bar{y}_{3..} = [3(\lambda_1 + \gamma_2) + 2(\lambda_1 + \gamma_3)]/5 - [2(\lambda_3 + \gamma_1) + 2(\lambda_3 + \gamma_2)]/4$$

$$= \lambda_1 - \lambda_3 - \frac{1}{2}\gamma_1 + \frac{1}{10}\gamma_2 + \frac{2}{5}\gamma_3 = \lambda_1 - \lambda_3 - 0.5\gamma_1 + 0.1\gamma_2 + 0.4\gamma_3$$

$$= 17.6 - 15.5 = 2.1,$$

whereas the least-square means for genotypes 1, 2, and 3 at the Q_1 locus are 15.40, 16.66 and 17.98, respectively, and the additive effect estimated by the difference between least-square means for genotypes 1 and 3 at the Q_1 locus is $15.40 - 17.98 = -2.59$. Remarkably, the ranks and magnitudes of the marginal and least-square means are reversed and the additive effects have different signs! The bias is $-2.59 - 2.10 = -4.69$, which is nearly two times the magnitude of the effect of the locus.

The favorable allele homozygote at the Q_1 locus is genotype 3, yet if the parameters of Q_1 had not been simultaneously estimated along with the parameters of Q_2, then it would be erroneously concluded that the favorable allele homozygote is genotype 1. If this information was subsequently used for marker-assisted selection, then negative selection pressure would be exerted against the favorable allele. In practice, of course, the sample size would be much greater than the sample we used for the example. We used it because it let us show the expected values and mean estimators expressed as a function of these values. Despite this, imagine this sort of uncertainty and bias translated across a genome segregating for many QTL.

This example illustrates how misleading nonsimultaneous hypothesis tests can be when the data are unbalanced. An estimate of the effect of one locus may be very misleading when other important QTL are excluded from the model. The only difference between the missing cell data (Table 5) and the original unbalanced data with no missing cells (Table 1) is two missing observations, yet there were dramatic differences in bias magnitude and direction. In addition, the additive effect of Q_1 estimated by the difference between least-square means for the original data (Table 1) was -2.61, which is nearly equal to the least-squares mean estimate (-2.59) from the missing cell data (Table 5); thus, while the marginal means fluctuated dramatically, the least-square means did not.

Full factorials of QTL genotypes do not exist for many multilocus models. To circumvent this problem, we have proposed handling multilocus QTL genotypes as one-dimensional arrays, instead of factorial arrays, i.e., using a means model [10]. This is just a mechanism for making it feasible to simultaneously estimate the parameters of multilocus main-effects models. By defining QTL genotypes as a one-dimensional array, there is just one factor for QTL genotypes, regardless of the number of QTL which are simultaneously mapped. This does not alleviate the problem of missing QTL genotypes, but it facilitates estimating the main effects of numerous QTL simultaneously. Hypotheses are simultaneously tested by using linear differences among the array of QTL genotypes. As we proved earlier, unbiased estimates of QTL genotype means and unbiased hypothesis tests exist as long as there are no missing QTL genotypes among the 2^k or 3^k genotypes for k independent loci or, if there are missing cells, no interactions between QTL genotypes (Fig. 1).

By using whatever array of multilocus QTL genotypes arising in a population, there are no missing cells, but there may be many missing genotypes among the 2^k or 3^k genotypes in a population. We have to be careful about the meaning of missing cells. Missing genotypes are hidden when using the means model. Even though there formally are no missing cells, the factorial structure and missing cells are still present.

For the unbalanced two-locus F_2 data without missing cells, for example, there are observations for every QTL genotype (Table 1). If the genotypes at these loci are taken as a one-dimensional array, then the model is reduced to one factor with nine QTL genotypes or factor levels. Using a one-dimensional array of QTL genotypes facilitates using the one-factor linear model

$$y_{ij} = \mu + \tau_i + e_{ij} \tag{7}$$

where $i = 1, 2, \ldots, 9$, $j = 1, 2, \ldots, n_i$, i indexes genotypes at quantitative trait loci Q_1 and Q_2, j indexes the number of observations of the ith QTL genotype, n_i is the number of observations of the ith QTL genotype, y_{ij} is the ijth observation of the quantitative trait, μ is the population mean, and τ_i is the effect of the ith genotype at Q_1 and Q_2. For the unbalanced two-locus F_2 data with missing cells, however, there are no observations for certain QTL genotypes (Table 5). Under model (6), there formally are no missing cells; however, the double homozygotes are clearly missing, and careful attention must be paid to this to avoid making misleading inferences.

Type I or false positive error probabilities

If an experimenter rejects a true null hypothesis, e.g., $H_0 : \mu_1 = \mu_3$ is rejected when it is true, then a Type I error has been made. The probability of a Type I or false positive error is a function of the stated probability of a Type I error and the number of true null hypotheses. Stated and achieved Type I error probabilities are equal only if every null hypothesis is true; thus, the stated probability is a conditional probability. The stated Type I error probability (α) predicts the correct number of false positives only when the null hypothesis is always true. If, for example, 100% of the null hypotheses are false, then there are no Type I errors and the stated and achieved Type I error probabilities are not equal. If, however, 100% of the null hypotheses are true, then the stated and achieved Type I error probabilities are equal.

The number of Type I errors increases as the number of dependent hypothesis tests increases. If QTL are separately or nonsimultaneously mapped, then tests of hypotheses about differences between QTL genotypes are not independent, and the number of Type I errors which arise is greater than the stated probability of Type I errors. To avoid this pitfall, Lander and Botstein [9] stressed using conservative Type I error probabilities. Although the objective of using conservative Type I error probabilities is to minimize false positives, it erodes power. It is preferable to relax the Type I error probability and maximize power if the objective is to map QTL for MAS because more QTL will be found, and selection of QTL with minor effects gain the most from MAS

[8]. Furthermore, the problem of false positives can be more efficiently minimized by using simultaneous hypothesis tests. This maximizes power and decreases the frequency of false negatives.

False positives are caused by random errors and sampling bias

Type I errors and power are usually defined by assuming the sole cause of Type I errors is random error. Type I errors are solely caused by random errors as long as the effects of QTL genotypes are not affected by nonrandom sources of error. As we showed earlier, if a hypothesis is tested about a QTL and the parameters of other important QTL are not simultaneously estimated, then the genotypes of QTL omitted from the model may bias estimates of parameters of QTL in the model. This bias is a source of nonrandom error.

In practice, bias is eliminated by simultaneously estimating the least-square means and Type III SS of important QTL within the limits we defined for missing cells and interactions between QTL (Fig. 1). A true null hypothesis might be rejected because the difference between means of QTL genotypes is positively biased. Likewise, a false null hypothesis might not be rejected because the difference between means of QTL genotypes is negatively biased. If there are QTL which are not linked to markers, then these are a source of bias which cannot be eliminated. These QTL should be envisioned as factors which can never be defined, but which exist.

An example: mapping QTL underlying phenotypic differences for telomere length among maize recombinant inbred lines

To illustrate the disparities which arise between nonsimultaneously and simultaneously mapped QTLs, we used example data from 41 maize recombinant inbred (RI) lines assayed for 70 RFLP loci. These data were kindly provided to us by Benjamin Burr of the Brookhaven National Laboratory to use to test the QTL-STAT software. The trait we examined was telomere length (kb), which was measured twice for each RI line. The experimental design was completely randomized; thus, a suitable linear model is

$$y_{ijk} = \mu + \lambda_i + \phi_{ij} + e_{ijk} \tag{8}$$

where $i = 1$ or 2, $j = 1, 2, \ldots, n_i$, $k = 1$ or 2, i indexes multilocus genotypes for the array of quantitative trait loci in the model, j indexes the number of lines of the ith QTL genotype, k indexes the number of observations of each recombinant inbred line, n_i is the number of observations of the ith QTL genotype, y_{ijk} is the ijkth observation of the quantitative trait, μ is the population mean, λ_i is the effect of the ith QTL genotype, and ϕ_{ij} is the effect of the jth recombinant inbred line nested in the ith QTL genotype. The effects of lines are random, whereas the effects of QTL genotypes are fixed (Table 6).

Table 6. Analysis of variance of a completely randomized experimental design for model (2). The treatments are QTL genotypes (QTL) and recombinant inbred lines nested in QTL genotypes. The effects of QTL genotypes are fixed. The effects of lines are random. Model (2) is suitable for an equal number of replications of lines and an unequal number of replications of QTL genotypes[a, b]

Source	Degrees of freedom
Line (L)	$df_n = N - 1$
QTL genotype (QTL)	$df_q = q - 1$
L:QTL	$df_{n:q} = N - q$
Residual	$df_e = N(r - 1)$

Expected mean square

$$E(M_n) = \sigma_e^2 + r\sigma_n^2 = \sigma_E^2 + r2\sigma_A^2 = \sigma_E^2 + r(2\sigma_A^2 - \phi_Q^2) + \left[\frac{(q-1)r\bar{n}}{(q\bar{n} - 1)}\right]\phi_Q^2$$

$$E(M_q) = \sigma_e^2 + r\sigma_{n:q}^2 + r\bar{n}\phi_Q^2 = \sigma_E^2 + r(2\sigma_A^2 - \phi_Q^2) + r\bar{n}\phi_Q^2$$

$$E(M_{n:q}) = \sigma_e^2 + r\sigma_{n:q}^2 = \sigma_E^2 + r(2\sigma_A^2 - \phi_Q^2)$$

$$E(M_e) = \sigma_e^2 = \sigma_E^2$$

[a] q is the number of QTL genotypes, $\bar{n} = \dfrac{\left(\sum\limits_{i=1}^{q} n_i^2\right) N}{q - 1}$, $N = \sum\limits_{i=1}^{q} n_i$, n_i is the number of lines of the *i*th QTL genotype or the number of replications of the *i*th QTL genotype, and r is the number of replications per line.

[b] σ_e^2 is the residual variance, $\sigma_{n:q}^2$ is the line nested in QTL genotype variance, σ_n^2 is the betweenline variance, ϕ_Q^2 are the fixed effects of QTL genotypes, σ_A^2 is the additive genetic variance, and σ_E^2 is the environmental variance. The causal expected mean squares were defined by assuming epistatic genetic variances were equal to zero.

QTL underlying phenotypic differences for telomere length were mapped using QTL-STAT. The null hypothesis of no QTL was initially tested for every 0.1 cM of the genome using LOD- and F-statistics. The LOD-statistic is

$$\text{LOD} = \log_{10}L = \frac{df_q F}{2\log(10)}$$

where L and $F = M_q/M_{n:q}$ are the likelihood ratio and F statistics, respectively, for testing the null hypothesis (Table 6); thus, LOD and L are linear functions of F when the y_{ijk} are normally distributed [17]. The mean square among lines nested in QTL genotypes ($M_{n:q}$) is used as the error for testing hypotheses about QTL genotype means since

$$\frac{E(M_q)}{E(M_{n:q})} = \frac{\sigma_e^2 + r\sigma_{n:q}^2 + r\bar{n}\phi_Q^2}{\sigma_e^2 + r\sigma_{n:q}^2} = 1 + \frac{r\bar{n}\phi_Q^2}{\sigma_e^2 + r\sigma_{n:q}^2}$$

(Table 6) [6].

LOD and F statistics were estimated for every 0.1 cM and used to make LOD score plots for each chromosome (Fig. 2 and 3). These plots are a continuous sequence of nonsimultaneous tests of the null hypothesis along each chromosome (Fig. 2 and 3). We found at least eight marker-brackets for which there was significant evidence for QTL (Table 7 and Fig. 2 and 3). The F statistics listed for these QTL are for the point where the LOD or F statistic was maximum within each of the eight marker-brackets (Table 7).

Because nonsimultaneous tests use Type I statistics, these tests are biased and must be interpreted with caution. To alleviate the bias problem, hypotheses about these QTL were simultaneously tested to determine which QTL should remain in the model. We iteratively re-estimated parameters and retested hypotheses until a stable model was found by simultaneous hypothesis tests. By stable we mean a model where every QTL explains a significant fraction of the phenotypic variance. The algorithm used by QTL-STAT to find a 'stable model' is equivalent to forward elmination. Other algorithms might be implemented and tested, but this is the only one we used for the telomere length example.

Table 7. Nonsimultaneous interval mapping by QTL-STAT of the genome of the Co159 × Tx303 maize recombinant inbred (RI) line population for quantitative trait loci (QTL) affecting telomere length (kb). Two replications of each RI line and a completely randomized experimental design were used. Degrees of freedom (df), Type I sums of squares (SS), and coefficients of determination (R^2) are listed for the additive effects of eight QTL, which are defined by marker-bracket names, e.g., *bnl*5.62–*bnl*8.05. The genetic variance between lines nested in QTL genotypes was used as the error for the F statistic (F) for testing the null hypothesis of no additive effect. The SS for each QTL and line nested in the QTL group adds up to the SS between lines because these are nonsimultaneous tests

Source	df	SS	Pr > F	R^2
Line	40	1,235.3	<0.0001	0.97
*bnl*5.62–*bnl*8.05	1	224.8	0.005	0.18
Line(*bnl*5.62–*bnl*8.05)	39	1,010.5		
*bnl*8.10–*ynh*20	1	243.1	0.004	0.19
Line(*bnl*8.10–*ynh*20)	39	992.2		
*zp*1*f*–*umc*31*a*	1	437.4	<0.0001	0.34
Line(*zp*1*f*–*umc*31*a*)	39	797.9		
*npi*208–*bnl*10.05	1	406.8	0.0001	0.32
Line(*npi*208–*bnl*10.05)	39	828.5		
*bnl*5.71–*bnl*10.12	1	211.9	0.007	0.17
Line(*bnl*5.71–*bnl*10.12)	39	1,023.4		
*bnl*8.21–*bnl*8.39	1	301.5	0.001	0.24
Line(*bnl*8.21–*bnl*8.39)	39	828.5		
*bnl*5.10–*acp*1	1	322.1	0.0006	0.25
Line(*bnl*5.10–*acp*1)	39	1,023.4		
*gln*1–*npi*350	1	709.7	<0.0001	0.56
Line(*gln*1–*npi*350)	39	525.6		
Residual	41	44.4		

230

Only three QTL remained in the final or stable model (Table 10). The model which preceded the final model had five QTL (Tables 8 and 9). Two QTL were dropped from this model because they failed to explain a significant fraction of the phenotypic variance (Tables 8 and 9); thus, a total of five QTL were eliminated because they did not have significant effects when tested simultaneously with the original set of eight QTL or a subset thereof. We repeatedly find that many of the QTL found by nonsimultaneous tests are not retained in the model after simultaneous tests. The four or five spurious QTL found by nonsimultaneous mapping were Type I errors (Tables 7, 8, 9, and 10).

The difference between the three- and five-locus models is very slight (Tables 8, 9, and 10). For the final model, the marker-brackets $bnl5.10-acp1$ and $npi208-bnl10.05$ were dropped because they did not explain a significant fraction of the genetic variance when tested simultaneously with the other three major loci (Tables 8 and 10). The marginal and least-square means and additive effects show why. Additive effects estimated by differences between marginal means for every QTL were inflated by sampling bias (Table 9). The magnitude of the bias ranged from 1.58 to 3.40 (Table 9). Additive effects estimated from marginal and least-square means for the $bnl5.10-acp1$ QTL, for example, were 3.98 and 0.58, respectively; thus, the additive effect of the $bnl5.10-acp1$ QTL was upwardly biased by 3.40 when the other four QTL were excluded from the model (Table 9). The $npi208-bnl10.05$ QTL was barely important enough to keep in the model, but it did account for roughly 3.0% of the phenotypic variance (Tables 8 and 9). If we presume there are no other important QTL and no interactions between QTL, then simultaneous estimation eliminated the bias from every QTL (Tables 7, 8, 9, and 10).

Table 8. Simultaneous interval mapping by QTL-STAT of the genome of the Co159 × Tx303 maize recombinant inbred (RI) line population for quantitative trait loci (QTL) affecting telomere length (kb). Two replications of each RI line and a completely randomized experimental design were used. Nonsimultaneous hypothesis tests were used to select QTL for simultaneous hypothesis tests. Degrees of freedom (df), Type III sums of squares (SS), and coefficients of determination (R^2) are listed for the additive effects of five simultaneously mapped QTL. The QTL are defined by marker-bracket names, e.g., $bnl5.62-bnl8.05$. The residual genetic variance between lines was used as the error for F statistics (F) for testing hypotheses about the additive effects of QTL

Source	df	SS	Pr > F	R^2
Line	40	1,235.3	<0.0001	0.97
QTL genotype	5	952.2	<0.0001	0.74
$bnl8.10-ynh20$	1	64.7	0.008	0.05
$zp1f-umc31a$	1	45.2	0.021	0.04
$npi208-bnl10.05$	1	33.9	0.048	0.03
$bnl5.10-acp1$	1	4.9	0.44	0.01
$gln1-npi350$	1	205.4	<0.0001	0.16
Line(QTL genotype)	35	283.1	<0.0001	0.22
Residual	41	44.4		

Table 9. Marginal and least-square means and additive effects for five QTL affecting telomere length (kb) in the Co159 × Tx303 maize recombinant inbred (RI) line population. The QTL are defined by marker-bracket names, e.g., *bnl*5.62–*bnl*8.05. The parameters of each QTL were nonsimultaneously and simultaneously estimated using QTL-STAT. Two replications of each RI line and a completely randomized experimental design were used

QTL	Genotype	Mean		Additive effect	
		Marginal	Least square	Marginal	Least square
*bnl*8.10–*ynh*20	11	8.11	8.86		
	22	11.64	10.81	– 3.53	– 1.95
*zp*1f–*umc*31a	11	12.75	10.75		
	22	8.12	8.93	4.63	1.82
*npi*208–*bnl*10.05	11	11.87	10.64		
	22	7.17	9.04	4.70	1.60
*bnl*5.10–*acp*1	11	12.40	10.13		
	22	8.42	9.55	3.98	0.58
*gln*1–*npi*350	11	7.25	7.93		
	22	13.14	11.74	– 5.89	– 3.81

There were major differences between SS estimates from nonsimultaneous and simultaneous hypothesis tests (Tables 7, 8, and 10). The sums of squares among QTL genotypes (S_q) for the *bnl*5.10–*acp*1 QTL were 322.1 and 4.9 for the individual locus and five locus models, respectively (Tables 7 and 8). This disparity was caused by sampling bias. Another way to show this, which is useful later, is to use the coefficient of determination for QTL genotypes (R_q^2) where $R_q^2 = S_q/S_t$ and S_t is the SS total. R_q^2 for the *bnl*5.10–*acp*1 QTL for individual and five-locus models were 0.25 and 0.01, respectively (Tables 7 and 8).

Table 10. Simultaneous interval mapping by QTL-STAT of the genome of the Co159 × Tx303 maize recombinant inbred (RI) line population for quantitative trait loci (QTL) affecting telomere length (kb). The QTL are defined by marker-bracket names, e.g., *bnl*5.62–*bnl*8.05. Two replications of each RI line and a completely randomized experimental design were used. Nonsimultaneous hypothesis tests were used to select QTL for simultaneous hypothesis tests. Degrees of freedom (df), Type III sums of squares (SS), and coefficients of determination (R^2) are listed for the additive effects of three simultaneously mapped QTL. The residual genetic variance between lines was used as the error for the F statistics (F) for testing hypotheses about the additive effects of QTL

Source	df	SS	Pr > F	R^2
Line	40	1,235.3	<0.0001	0.97
QTL genotype	3	914.0	<0.0001	0.71
*bnl*8.10–*ynh*20	1	88.2	0.003	0.07
*zp*1f–*umc*31a	1	104.5	0.001	0.08
*gln*1–*npi*350	1	313.9	<0.0001	0.25
Line(QTL genotype)	37	321.3	<0.0001	0.25
Residual	41	44.4		

Naturally, the SS between lines nested within QTL genotypes ($S_{n:q}$) were less for the three- and five-locus models (Tables 8 and 10) than for the individual locus models (Table 7). This is important because $M_{n:q} = \hat{\sigma}_e^2 + r\hat{\sigma}_{n:q}^2$ is the error used to test hypotheses about QTL effects, and thus partly determines power. Intuitively, power is expected to be greater for simultaneous than for nonsimultaneous hypothesis tests [6]. This holds for unbiased hypothesis tests, but not necessarily for biased hypothesis tests. While the error variances for multilocus models are often less than for individual locus models, significant power increases may not be achieved because the variances for QTL are often inflated by bias; thus, while simultaneous mapping often decreases the error variance, it often decreases the variances for QTL as well. The error variances for nonsimultaneous tests of the hypothesis of no QTL within the *bnl*8.10–*ynh*20 and *gln*1–*npi*350 marker-brackets were 25.4 and 13.5 (Table 7), respectively, whereas the error variance for the three-locus model was 8.7 (Table 10). But, M_q for *bnl*8.10–*ynh*20 for the three-locus model was significantly less than for the individual locus model, viz. 88.2 as opposed to 243.1 (Tables 7 and 10). Ironically then many loci are dropped after simultaneous tests even though the error is greatly decreased. This seems counterintuitive, but power is not necessarily increased by simultaneous tests, as we have previously argued [6], because sampling bias often inflates the power of nonsimultaneous tests. This factor was overlooked when we originally projected power increases for simultaneous tests [6].

Replicated progeny are useful for estimating classical quantitative genetic parameters, in addition to QTL parameters, e.g., the additive genetic variance between lines and line-mean heritability. It is useful to estimate these parameters to gain a more thorough understanding of a trait and the QTL which underlie it.

As we showed earlier, R_q^2 must be carefully interpreted when the data are unbalanced. If the R_q^2 for a QTL model is 15%, then this should not be interpreted to mean that the model is inadequate because the model might explain 100% of the genetic variance. The coefficient of determination between lines ($R_n^2 = S_n/S_t$) can be used to monitor this (Table 6).

If $R_q^2 = 0.15$, then it might be erroneously concluded that the QTL model does not adequately explain the genetics of the trait because it only explains 15% of the phenotypic variance. But if $R_n^2 = 0.15$, then the QTL model explains 100% of the genetic variance between lines. This is clear from the expected mean squares for model (7). The SS between lines (S_n) is equal to the sum of the SS between QTL genotypes (S_q) and the SS between lines nested within QTL genotypes ($S_{n:q}$) ([6] and Table 6). $S_{n:q}$ is the between-line SS which is not explained by simultaneously estimated QTL. 100% of the genetic variance between lines is explained by QTL when $R_n^2 = R_q^2$ or, equivalently, when $S_n = S_q$; thus, a useful estimate of the fraction of the genetic variation explained by a QTL model is S_q/S_n. This ratio ranges from 0.00 to 1.00 and is only estimable from replicated progeny data. As before, this statistic must be carefully interpreted.

We listed the Type III SS for individual QTL and for the full QTL model for the three- and five-locus models (Tables 8 and 10). The Type III SS and R_q^2 for individual QTL effects within multilocus models are not additive. The sum of the Type III SS for the three QTL in the final model, for example, does not add up to the SS for the full QTL model $(88.2 + 104.5 + 313.9 \neq 914.0)$; likewise, the sum of the R_q^2 for individual QTL do not add up to the R_q^2 for the full QTL model $(0.07 + 0.08 + 0.25 \neq 0.71)$ (Table 10); hence, using R_q^2 for an individual QTL can be misleading. For the maize example, the individual R_q^2 were much less than R_q^2 for the full model (Tables 8 and 10).

The line-mean heritability for telomere length is

$$ H_{RI} = \frac{\hat{\sigma}_n^2}{\hat{\sigma}_p^2} = \frac{\hat{\sigma}_n^2}{\hat{\sigma}_e^2/r + \hat{\sigma}_n^2} = \frac{2\hat{\sigma}_A^2}{\hat{\sigma}_E^2/r + 2\hat{\sigma}_A^2} = 1 - \frac{M_e}{M_n} = 1 - \frac{1.083}{30.883} = 0.965 . $$

where $\hat{\sigma}_p^2$ is the line-mean phenotypic variance (Tables 6 and 7). Although the heritability for telomere length is close to 1.0, the phenotypic distribution for this trait is continuous because many genes or QTL underlie it. The genetic variance between RI lines is twice the additive genetic variance, so an estimate of the additive genetic variance for telomere length is

$$ \hat{\sigma}_A^2 = \frac{\hat{\sigma}_n^2}{2} = \frac{M_n - M_e}{2r} = \frac{30.88 - 1.08}{4} = 7.45 $$

(Tables 6 and 7).

It is tempting to express the variance explained by QTL as a function of the additive genetic variance or the line-mean phenotypic variance. To do this, we forced the QTL genotypes to be random effects and defined σ_q^2 as the variance among QTL genotypes, then

$$ \frac{\sigma_q^2}{\sigma_n^2} = \frac{\sigma_q^2}{2\sigma_A^2} = \frac{[E(M_q) - E(M_{n:q})]/(r\bar{n})}{[E(M_n) - E(M_e)]/r} $$

and

$$ \frac{\sigma_q^2}{\sigma_p^2} = \frac{\sigma_q^2}{\sigma_e^2/r + \sigma_n^2} = \frac{\sigma_q^2}{\sigma_E^2/r + 2\sigma_A^2} = \frac{[E(M_q) - E(M_{n:q})]/(r\bar{n})}{[E(M_n)]/r} . $$

(Table 6). σ_q^2/σ_n^2 converges to 1.0 and σ_q^2/σ_p^2 converges to H_{RI} sometimes, but σ_q^2/σ_n^2 may have values greater than 1.0 and σ_q^2/σ_p^2 may have values greater than H_{RI}. This is illustrated for different models of QTL affecting telomere length (Table 11). The variance among $gln1-npi350$ genotypes, for example, was 16.9, which is greater than the variance among RI lines (Table 11) because this locus has a major effect. Likewise, the genetic variance among genotypes for the three major QTL exceeds the genetic variance among lines but is less than the variance among $gln1-npi350$ genotypes (Table 11).

Table 11. Variances among RI lines and QTL genotypes and their ratios with the line-mean phenotypic variance (σ^2/σ_p^2) and coefficients of determination (R^2) for different QTL models for telomere length. QTL genotypes were estimated using QTL-STAT

Parameter[1]	Lines	QTL[2]	QTL[3]	QTL[4]
σ^2	14.9	14.2	14.8	16.9
σ^2/σ_p^2	0.97	0.91	0.96	1.10
R^2	0.97	0.90	0.72	0.55

[1] σ^2 is either the genetic variance among RI lines (σ_n^2) or the genetic variance among QTL genotypes (σ_q^2) for a specific QTL model. R is either R_n^2 or R_q^2.
[2] This model used QTL genotypes for the *bnl*8.21–*bnl*8.39, *bnl*5.62–*bnl*8.05, *bnl*5.71–*bnl*10.12, *bnl*5.10–*acp*1, *npi*208–*bnl*10.05, *bnl*8.10–*ynh*20, *zp*1*f*–*umc*31*a*, and *gln*1–*npi*350 marker-brackets as the random effects. These QTL made up the group which had the greatest effects when tested nonsimultaneously (Table 7).
[3] This model used QTL genotypes for the *bnl*8.10–*ynh*20, *zp*1*f*–*umc*31*a*, and *gln*1–*npi*350 marker-brackets as the random effects. These QTL made up the stable model (Table 10).
[4] This model used genotypes for the *gln*1–*npi*350 marker-bracket as the random effects. This QTL had the greatest effect in every model.

The genetic variance among genotypes for the eight original QTL was closest to the between-line variance (Table 11). Furthermore, the SS among these eight QTL explained $0.90/0.97 = 0.93$ or 93.0% of the genetic variation for telomere length; however, many of these were dropped from the model because they failed to explain a significant fraction of the phenotypic variance when tested simultaneously. This is an important point which we have made before [6]. If the number of QTL genotypes is equal to the number of lines, as would happen for models with many loci mapped simultaneously, then $\sigma_n^2 = \sigma_q^2$, but this does not mean every important QTL has been found. This happens when the number of QTL increases because the multilocus genotype of every line is going to be unique when you examine enough loci simultaneously; thus, R_q^2 alone is not a reliable measure of the adequacy of a model.

Line-mean heritability is useful for estimating the effect of selecting among lines. This is the only estimator which makes sense for RI lines because there is no within-line genetic variance (Table 6). For the telomere length example, H_{RI} is equal to

$$R_n^2 = \frac{1235.4}{1279.7} = 0.965,$$

but $H_{RI} = R_n^2$ only when heritability is close to 1.00.

R_q^2 for individual QTL from nonsimultaneous hypothesis tests ranged from 0.17 to 0.56 (Table 7), which is close to the range of S_q/S_n because R_q^2 is close to 1.00. R_q^2 estimated using Type I SS for individual QTL models (Table 7) are obviously not additive and, because Type I SS are often seriously biased, may be misleading. The R_q^2 for individual QTL which are part of the three-locus model were 0.19, 0.34, and 0.56 (Table 7). The sum of these statistics is 1.09. These QTL obviously do not explain 109.0% of the phenotypic variation. R_q^2

for the three- and five-locus models were 0.71 (Table 10) and 0.74 (Table 8), respectively; thus, these models explain 76.4% and 73.0% of the genetic variation between lines, respectively. Most of the genetic variation for telomere length was explained by three or four QTL, but these QTL still fail to explain roughly 25.0% of the genetic variation for telomere length. Some of this unexplained genetic variation might be explained by epistasis and false negative QTL.

The method we used to put the multilocus model together does not overcome the problem of false negatives. This is substantiated by the direction of the bias we uncovered (Tables 7, 8, 9, and 10). Every nonsimultaneous test was biased upward (Tables 7, 8, 9, and 10). This was because initial mapping was done using individual marker-brackets. As we showed earlier, an effective way to find and eliminate false positives is by using simultaneous hypothesis tests. We have no way of knowing for certain if important QTL were excluded from the model for telomere length QTL as a consequence of downward bias, but given that 25.0% of the genetic variation is not explained by the three-locus model, it is probable that there are false negatives. Remapping the genome with the effects of important QTL estimated simultaneously might uncover QTL which were missed by nonsimultaneous mapping. An algorithm which iteratively maps the genome using multilocus methods has not yet been implemented, but it is a worthwhile objective. Finally, the theory and examples show that finding important QTL is straightforward, but finding and mapping every important QTL simultaneously is not quite as straightforward.

Acknowledgements

We are grateful to Ben Burr for kindly sharing the telomere length data with us and to Marlin Edwards, Brian Yandell, Albrecht Melchinger, and Chris Schon for criticisms which significantly improved the manuscript. Oregon Agricultural Experiment Station Technical Paper No. 9631.

237

References

1. Beckmann JS, Soller M (1986) Restriction fragment length polymorphisms in plant genetic improvement. Oxford Surv Plant Mol Biol 3: 197–246.
2. Edwards MD, Stuber CW, Wendel JF (1987) Molecular marker facilitated investigations of quantitative trait loci in maize. I. Numbers, genomic distribution, and types of gene action. Genetics 116: 113–125.
3. Fruend RJ, Littell RC, Spector PC (1986) SAS® system for linear models. SAS, Cary, North Carolina.
4. Jensen J (1989) Estimation of recombination parameters between a quantitative trait locus (QTL) and two marker gene loci. Theor Appl Genet 78: 613–618.
5. Knapp SJ, Bridges WC, Birkes D (1990) Mapping quantitative trait loci using molecular marker linkage maps. Theor Appl Genet 79: 583–592.
6. Knapp SJ, Bridges WC (1990) Using molecular markers to estimate quantitative genetic parameters: Power and genetic variances for unreplicated and replicated progeny. Genetics 126: 769–777.
7. Knapp SJ (1991) Using molecular markers to map multiple quantitative trait loci: Models for backcross, recombinant inbred, and doubled haploid progeny. Theor Appl Genet 81: 333–338.
8. Lande R, Thompson R (1990) Efficiency of marker-assisted selection in the improvement of quantitative traits. Genetics 124: 743–756.
9. Lander ES, Botstein D (1989) Mapping Mendelian factors underlying quantitative traits using RFLP linkage maps. Genetics 121: 185–199.
10. Milliken GA, Johnson DE (1984) Analysis of Messy Data. Volume I: Designed experiments. Van Nostrand Reinhold, New York.
11. Searle S (1971) Linear Models. New York: John Wiley.
12. Simpson SP (1989) Detection of linkage between quantitative trait loci and restriction fragment length polymorphisms using inbred lines. Theor Appl Genet 77: 815–819.
13. Soller M, Beckmann JS (1990) Marker-based mapping of quantitative trait loci using replicated progenies. Theor Appl Genet 79: 205–208.
14. Tanksley SD, Young ND, Paterson AH, Bonierbale MW (1989) RFLP mapping in plant breeding: new tools for an old science. Bio/technology 7: 257–264.
15. Williams JGK, Kubelik AR, Livak KJ, Rafalski JA, Tingey SV (1990) DNA polymorphisms amplified by arbitrary primers are useful as genetic markers. Nucleic Acids Res 18: 6531–6535.
16. Weller JI (1986) Maximum likelihood techniques for the mapping and analysis of quantitative trait loci with the aid of genetic markers. Biometrics 42: 627–640.
17. Yandell BS (1991) Quantitative trait loci in *Brassica rapa*. In: Keramidas E (ed), Proc. 23rd Symp. Interface.
18. Young ND, Tanksley SD (1989) Restriction fragment length polymorphism maps and the concept of graphical genotypes. Theor Appl Genet 77: 95–101.
19. Young ND, Tanksley SD (1989) RFLP analysis of the size of chromosomal segments retained around the *Tm-2* locus of tomato during backcross breeding. Theor Appl Genet 77: 95–101.

Epilogue: additional reflections

JACQUES S. BECKMANN

Mapping, for all practical purposes, is a stoichastic and cumulative process, with more and more 'arbitrary' tags, the mapped marker loci (these are certainly arbitrary with respect to their genetic location), added onto the map. The usual scenario goes as follows: a random – often anonymous – DNA segment will be attributed to a specific genetic location. Both the probe and the identified locus will be given some identifying coordinates. Provided that this probe is distributed or its sequence known, access to the corresponding genetic region will be public. If this probe also hybridizes to heterologous species, it can also be utilized to generate genetic data for a second species.

Must each mapping task be a new effort?

At present this is often the case, as one species map – based in large part on random anonymous DNA markers – will have been generated in one laboratory, while another map of the same species, showing limited genetic overlap with the first one, will have independently seen the light in a second lab. This was the situation for maize. Fortunately, the markers were placed, thanks to a collaborative effort, onto the same recombinant inbred map [3]. Such an elegant solution is, however, not always available for lack of recombinant inbred lines. The difficulties in integrating the different maps are further compounded by the fact that one also needs to take into account the influence the nature of the cross (whether it involves widely divergent or closely related members of the species) is likely to have on the observed recombination values.

An analogous situation prevails when comparing maps between species. Here too a set of random anonymous DNA probes will frequently have been utilized to generate data on a first species, a second set on the other species, etc. Yet, this time a comparison of the maps is further compounded with the inability to cross these species.

These situations, wherein – as a result of disconcerted efforts – distinct maps exist for one species or where each species is approached *de novo*, might bestow a considerable loss of information. Loss resulting from the existence of non-overlapping genetic maps is easy to visualize. Indeed, in practical life, one might often want to combine, for a given species, data from different independent reference maps in order to get adequate genome coverage in a particular chromosomal region. As a result, however, of the limited amount of overlap of these maps (due to a lack of common reference points), relating them one to another might frequently be equivalent to starting a new mapping exercise.

239

J.S. Beckmann and T.C. Osborn (eds) Plant Genomes: Methods for Genetic and Physical Mapping, 239–245.

Furthermore, loss imparted by the species-specific approach results from the fact that no benefit can be realized from the prior knowledge of linkage relationships in related species. This anticipated added value stems from the observation that loci which are closely linked to one another in one species stand a high likelihood of maintaining this privileged relationship in other species as well, and particularly in closely related ones. This property, termed conserved syntenies, i.e., the conservation of syntenic groups in more than one species, is currently used by gene mappers as a process to speed up and integrate mapping across all mammalian species. As a matter of fact, an entire chapter devoted to comparative mapping is included in the 10th workshop on Human Gene Mapping [4].

Expanding the concept of synteny conservation to the entire genome implies that it might suffice to have a good 'reference' map for one species, in order to have, at first approximation, the 'local' map of most other cultivated plants. We hope the reader will show indulgence to the liberty taken here in making this big a jump across kingdoms – maize and tomato are much further apart, in evolutionary terms, than mouse and man. And, it is still totally unknown how much these two plants share in genomic organization. Yet, existing evidence (see for instance the extensive syntenic conservation between the tomato and potato genomes [2]) suggests that this principle could apply at least to closely related plant species. In this case, it is evident that comparative mapping could be of great utility to speed up genetic coverage of a map. Furthermore, as a result of the wide variability of the ratios of physical to genetic distances in different plant species, this could prove to be an extremely valuable tool in attempts to clone specific genes by means of reverse genetics.

For this to happen, however, maps need to be interpretable across species, i.e., reference points must be transposable. One way to achieve this is by using as reference points defined evolutionary conserved landmarks (the best ones being the actual genes), rather than or on top of anonymous probes. Extending this idea further, eventually one important result of all this genetic mapping effort will be the complete catalogue of all genes and their genetic coordinates on the map, i.e., essentially a gene map.

Superposing a gene map onto the genetic map

To generate gene maps for different species, mapping efforts will have to be repeated, at least as many times as there are species of interest. Given the gene coordinates, it will be possible to resort to comparative mapping (and this independently of whether or not heterologous probes can be utilized – what matters is that homologous genes be examined). In other words, it will then be possible to compare two maps (both within and between species) and locate the breakpoints, i.e. those bifurcation sites where one maps departs from another (onto another evolutionary route). It is thus not difficult to envisage that one should be able to eventually translate these maps into a set of common reference maps that would have the gene orders as well as the species-specific translocation/rearrangement coordinates.

To illustrate this point, just consider the benefits that are to be gained from the ongoing genetic and physical mapping of *Arabidopsis*. It should not be too long before a large number of coding sequences are identified and mapped in this plant. This knowledge could, we suspect, readily be applied to genetic programs on the Cruciferae. And the ramifications use of this knowledge may have on applications of reverse genetics are easy to envisage.

What are the chances of identifying a useful genetic polymorphism in association with a defined gene?

RFLPs uncovered with cDNA probes certainly meet this criterion. It is also possible that undermethylated genomic DNA sequences, which are frequently used as a source of single-copy RFLP probes, may be enriched for coding sequences. But there are other means to gain access to potentially highly informative markers in the vicinity of genes. To address this point, it is worth making a short parenthesis in the animal kingdom and note that microsatellites – with the *proviso* that the applicability of microsatellite-associated poly- morphism to genetic studies in plants be demonstrated – seem to represent a very adequate type of marker for this kind of exercise. As a reminder, micro- satellites are short sequences comprised of tandemly repeated copies of a simple sequence motif (say a mono-, di-, tri- or tetranucleotide unit). These sites seem to be ubiquitous throughout the genome, and are often found to be associated with a length polymorphism which is due to a variation in the copy number of the repeated unit [e.g. 5, 8]. This polymorphism is visualized upon DNA amplification by the polymerase chain reaction, using site-specific oligonucleotide primers. Studies on mammals have shown that such sequence tagged microsatellite sites (STMS) can be expected to be present on the average every 50,000 base pairs. A number of human and rodent STMS markers have already been developed. These are often found to be multiallelic, and it is anticipated that PCR-formatted STMS will play a central role in human genetics. In addition, as these STMS are often associated with or in close proximity to structural genes, they can thus serve as excellent landmarks for such loci [9]. In other words, an immediate consequence of this distribution is that in the vicinity of almost any 'gene', even if the latter is essentially monomorphic, lies potentially such a useful polymorphic marker [1]. Provided the observations on the frequent intimate association of STMS with structural genes are extended to plant genomes, it should be possible to switch from a map of arbitrary anonymous genetic markers, to one representing essentially a genetic 'gene map'.

242

On genes and genomes

How many reference maps will eventually be needed? No one knows. It is far too optimistic and probably unrealistic to expect that one or even a few will do. Indeed, the number of species of interest is so great, and these often evolutionarily too distant, that defining a common base(s) for all plants, although it may be central, might be quite difficult a task if not impossible. But even so, one can still proceed by steps, generating at first a small number of reference maps (e.g. one for the Solanaceae, one for the Cruciferae, etc) and with time, enough insight will be gained to challenge the proposal of further reducing this number (e.g. deriving a map for a dicot, say *Arabidopsis*, for which the elaboration of a physical map is already near completion, one for a monocot, say maize).

But having genes and translocation points coordinates is still for the future. Nevertheless, it might be advisable now already to ensure adequate communication channels between the different species maps. As is clear from the preceding discussion, one such means would be that maps all share a set of standard reference points useful for cross-species analyses, defining 'genetic bands'. Intermediate loci (some of which will fall on known genes, other on arbitrary DNA segments) will be positioned, within each band, with respect to these reference points. (Another advantage of this reference system is that the reference markers are provisional, i.e., the width of the 'genetic bands' can be modulated to meet the demands imposed by progress or the required resolution power: the standard map could be made to include more reference points, as to provide a more detailed map. As new candidate landmarks will be uncovered in the course of this work, these can be added onto the map as required.)

A harmonization of effort is required

It seems therefore the more urgent, for these maps to be of optimal utility, both across labs as well as across plant species, to minimize inconsistencies, to discourage investigators to devise separate nomenclatures for each specialized genetic system. This should be accompanied by a uniformization of the procedure as well as by the adoption of a common basis for the naming of homologous mapped plant genes [e.g. 7]. Alternatively, adequate links need to be established between the different genetic databases to ensure an easy translation of genetic data from one map onto another.

This proposal has immediate practical implications. Let us consider the existing *Arabidopsis,* maize, tomato, potato or petunia genetic maps. These currently comprise a mixture of sites, a proportion of which might belong to the category of 'phylogenetically conserved landmarks'. (In our context, the term 'landmark' is often used to refer to a polymorphic marker, yet it should be remembered that it fulfills several important applications, as physical or genetic tag, whether or not it itself is polymorphic.)

These evolutionary conserved tags could be given in their nomenclature, a symbol allowing for their recognition as a function of their quality (e.g. degree of conservation), and the corresponding probes (or sequence tagged site specifics [6]) would be made publicly available. A consensus map could thus be derived for each one of these species. Eventually these different maps will use the same reference points. In the process of constructing maps for other species, the corresponding homologous genes would be the first candidates to be mapped.

And this may be the time to start

In other words, there is no need to wait until complete mapping has been achieved for a number of species. Quite the contrary, it would be worthwhile for current mapping to include attempts to integrate this same procedure onto the building of so-called reference map(s). Such map(s) would have the merit of allowing the comparisons of syntenic conservations over (short) chromosomal segments accross plant species.

Comparative mapping should prove useful in the elaboration of saturated maps. The maps will evolve as more and more data accumulate, upon incorporation of all the corrections. This will gradually and constantly lead to more refined maps. Altogether this should provide a powerful tool to facilitate (in a synergistic manner) the mapping of the plant genomes. This has practical applications as well, as the principle of synteny conservation could further be extended to loci associated with particular phenotypes. As already mentioned earlier, this could represent a potentially very powerful tool for the cloning of specific genes through 'species-hopping'. In addition, these maps would also yield invaluable information, on the evolution of plant genomes as well as on the evolution of defined characters.

To summarize, to benefit from the synergism bestowed by synteny conservation, i.e., to be able to extrapolate from one plant map to another, there is an urgent necessity that a common language be utilized for all maps, or in other words, that the genetic databases be interpretable by all plant mappers. This could be achieved by using non-anonymous DNA segments that are conserved throughout the plant kingdom (e.g. a known gene or cDNA sequence – leading eventually to a 'gene map') as landmark. It would be useful if all nomenclature inconsistencies be eliminated. This is the basis for the present call for an effort to rationalize procedures and standardize nomenclature of genetic locations, so that gene symbols for one species (or lab) won't conflict with gene symbols from another species (or lab).

The preceding ideas are certainly not intended as a proposal to this end. They were all given here as examples of issues that need to be dealt with, as ideas for future consideration. Everything, at this stage, is wide open. But time is ripe now for the scientific community to provide the coherent and adequate guidelines toward this end.

One last word

Our objective is not only to map genomes, but to provide eventually the means for a practical and rational implementation of this knowledge in breeding. It is thus important to restate that molecular geneticists do not face this genetic challenge alone. Plant geneticists and breeders have an important role to assume jointly in this endeavor. Substantial benefits should be made by helping the molecular geneticists to optimize strategies, by starting now already, using all the versatility and plasticity of plant genetics, to build the most convenient genetic stocks that will be used later – once the maps (or markers) will be ready – for the analyses of traits of interest.

245

References

1. Beckmann JS, Soller M (1990) Toward a unified approach to genetic mapping of eukaryotes based on sequence tagged microsatellite sites. Bio/technology 8: 930–932.
2. Bonierbale MW, Plaisted, RL, Tanksley SD (1988) RFLP maps based on a common set of clones reveal modes of chromosomal evolution in potato and tomato. Genetics 16: 91–106.
3. Burr B, Burr FA (1991) Recombinant inbreds for molecular mapping in maize: theoretical and practical considerations. Trends Genet 7: 55–60.
4. Lalley PA *et al.* (1989) Report of the committee on comparative mapping. Cytogenet Cell Genet 51: 503–532.
5. Litt M, Luty JA (1989) A hypervariable microsatellite revealed by *in vitro* amplification of a dinucleotide repeat within the cardiac muscle actin gene. Am J Hum Genet 44: 397–401.
6. Olson M, Hood L, Cantor C, Botstein D (1989) A common language for physical mapping of the human genome. Science 245: 1434–1435.
7. Price CA (1989) Nomenclature for cloned plant genes. Plant Mol Biol Rep 7: 99–103.
8. Weber JL, May PE (1989) Abundant class of human DNA polymorphisms which can be typed using the polymerase chain reaction. Am J Hum Genet 44: 388–396.
9. Williamson R *et al.* Report of the DNA Committee and Catalogues of Cloned and Mapped Genes, Markers Formatted for PCR and DNA Polymorphisms. Human Gene Mapping Workshop 11 (In press).

Index

248